Synthesis Lectures on Chemical Engineering and Biochemical Engineering

Series Editor

Bob Beitle Jr., College of Engineering, University of Arkansas at Fayetteville, Fayetteville, AR, USA

This series publishes short books on all aspects of chemical engineering, covering the analysis or design of chemical processes to effectively convert materials into more useful materials or energy. The books will focus on fundamental aspects necessary for chemical engineering design including chemistry, math, physics, and sometimes biology to improve the quality of life by inventing, optimizing, and economizing new technologies and products.

Guillermo Fernando Barreto ·
Carlos Daniel Luzi

Applying Multiple-Reaction Stoichiometry to Chemical Reactor Modelling

Springer

Guillermo Fernando Barreto
Departamento de Ingeniería Química
Facultad de Ingeniería
Universidad Nacional de La Plata
La Plata, Argentina

Carlos Daniel Luzi
Consejo Nacional de Investigaciones
Científicas y Técnicas (CONICET)
Buenos Aires, Argentina

ISSN 2327-6738 ISSN 2327-6746 (electronic)
Synthesis Lectures on Chemical Engineering and Biochemical Engineering
ISBN 978-3-031-42377-2 ISBN 978-3-031-42375-8 (eBook)
https://doi.org/10.1007/978-3-031-42375-8

This Springer imprint is published by the registered company Springer Nature Switzerland AG
The registered company address is: Gewerbestrasse 11, 6330 Cham, Switzerland

Preface

For a single reaction, net production rates r_j of the reacting species are simply related between them by the proportionality of their respective stoichiometric coefficients. The evolution of the system composition then admits a single degree of freedom and can be evaluated from a single mass balance of a given species. This fact, which is invariably invoked in introductory courses of Chemical Reactions Engineering (CRE), facilitates the treatment of (ideal) chemical reactors, either with time evolution ("batch" reactors) or spatial evolution ("plug-flow" reactors).

Conversely, the treatment of multiple reactions in introductory courses and CRE textbooks is not uniform. Although stoichiometry allows relating the net production rates r_j in a totally general way and thus extending their application to the reaction systems referred to in the previous paragraph, such relationships are very often not considered. In the authors' opinion, this is due to the fact that they do not directly arise from a given reaction set, but a previous algebraic treatment is required.

One of the motivations of this text is to review the alternatives for specifying the stoichiometric relationships between the net production rates r_j and to propose a satisfactory alternative for introductory CRE courses. Chapter 1 mainly pursues this objective.

Recognising the existence of such relationships allows the student to visualise the scope of stoichiometry in a completely general way, covering both homogeneous and heterogeneous reaction systems and, in addition, their application to the elementary steps of reaction mechanisms.

Specifically to the teaching of CRE, introductory CRE courses typically cover the study of basic chemical reactors based on plug-flow or perfect mixing hypothesis. Under such ideal assumptions and for continuous-flow or batch reactors, the relationships between the net production rates r_j allow establishing equivalent relationships between the variations of the species concentration, either locally or temporally. Instead, some limitations arise in open batch operations (semi-batch reactors). In this way, for most reacting systems treated in introductory CRE courses, the number of degrees of freedom that set the system composition becomes established by the stoichiometry, and furthermore the solution of the mass balances is simplified. All these aspects are addressed in Chap. 2, together with a discussion about the choice of suitable measures of the species concentration for this purpose.

Chapter 3 explores the possibility of extending the stoichiometric relationships between the net production rates r_j to equivalent relationships between the variations of the species concentration in systems with higher complexity than those discussed in Chap. 2. Distributed flow systems in more than one spatial direction, as in homogeneous and catalytic packed-bed reactors, are analysed by considering effects such as dispersion, unsteady state and presence of permeable walls (e.g., membrane walls). Restrictions are identified for situations of practical interest that do allow extending the stoichiometric relationships to variations of the species concentration. Possible advantages of using the stoichiometric relationships for the solution procedure of the mass conservation equations are also discussed.

The material in the present text is primarily intended for instructors of CRE courses. In particular, Sect. 1.4 of Chap. 1 details the procedure for specifying the stoichiometric relationships between the net production rates r_j, which can also be suggested as a suitable reading for students attending an introductory CRE course. Besides, it is worth mentioning that Chaps. 2 and 3 go beyond the sole concept of exploiting the stoichiometric relationship between the net production rates. Thus, Chap. 2 also provides a summary of several issues typically covered in the teaching practice of the different homogeneous reactors with ideal-flow hypotheses. Chapter 3 goes further and briefly discusses more advanced chemical reactor models employed in the CRE bibliography, some common limitations and simplifications, and dependencies of model parameters on the state variables. It is then expected that the content in Chap. 3 can be valuable for readers interested in CRE topics in general.

La Plata, Argentina — Guillermo Fernando Barreto
Buenos Aires, Argentina — Carlos Daniel Luzi

Contents

Nomenclature

A_j	Chemical species identified by index j
a_p	Surface-area to volume ratio of the pellet (m^{-1})
a_w	Perimeter per unit cross-section area (m^{-1})
C	Total molar concentration (mol m^{-3})
C_j	Molar concentration of A_j (mol m^{-3})
$\hat{c}_p$	Specific heat (J kg^{-1} K^{-1})
$\tilde{c}_p$	Molar heat capacity (J mol^{-1} K^{-1})
$\hat{c}_{pj}$	Specific heat of A_j in the mixture (J kg^{-1} K^{-1})
$\mathbb{C}_j$	Component variable associated with the molar concentration of A_j (mol m^{-3})
$\check{C}_j = F_j/q_0$	Relative variable respect to the inlet volumetric flow rate (mol m^{-3})
D_j	Axial or effective dispersion coefficient of A_j (m^2 s^{-1})
$D_{j,conv}$	Contribution to D_j from flow effects in packed bed (m^2 s^{-1})
$\mathfrak{D}_j$	Molecular diffusion coefficient of A_j in the mixture (m^2 s^{-1})
$\mathfrak{D}_{j,i}$	Binary diffusion coefficient between A_j and A_i (m^2s^{-1})
$\mathfrak{D}_{j,ef}$	Effective diffusion coefficient of A_j in a porous medium (m^2 s^{-1})
d_p	Pellet diameter (m)
d_t	Hydraulic diameter of the tube (m)
E	Total number of chemical elements present in S species (Chap. 1); number of species undergoing significant interfacial transfer (Chap. 3)
$E(t)$	Integrating factor (Chap. 2)
F	Total molar flow rate (mol s^{-1})
F_j	Molar flow rate of A_j (mol s^{-1})
f_j	Fugacity of A_j (atm)
$\mathbb{F}_j$	Component variable associated with the molar flow rate of A_j (mol s^{-1})
G	Mass velocity (kg m^{-2} s^{-1})

G_j	Mass flux of A_j (kg m^{-2} s^{-1})
G_{jw}	Interfacial mass flux of A_j through the solid boundaries (kg m^{-2} s^{-1})
G_{jp}	Interfacial (fluid/pellet) mass flux of A_j (kg m^{-2} s^{-1})
$\mathbb{G}_{jw}$	Component variable associated with G_{jw}, Eq. (3.22c) in Chap. 3 (kg m^{-2} s^{-1})
G_w	Total interfacial mass flux through the solid boundaries (kg m^{-2} s^{-1})
G_p	Total interfacial (fluid/pellet) mass flux (kg m^{-2} s^{-1})
$\hat{h}$	Specific enthalpy (J kg^{-1})
$\hat{h}_j$	Specific enthalpy of A_j (J kg^{-1})
h_j	Partial molar enthalpy of A_j (J mol^{-1})
I	Identity matrix (Chap. 1); identity tensor (Chap. 3)
$\mathbb{J}_j$	Component variable associated with $(\mathfrak{D}_{j,ef}\omega_j)$ (m^2 s^{-1})
$J_k = (-\Delta H_k)/(\rho\hat{c}_p)$	Adiabatic temperature rise per unit molar concentration of the kth key species (K m^3 mol^{-1})
j_j	Dispersion or diffusion mass flux of A_j (kg m^{-2} s^{-1})
K	Number of key species
K_{RI}	Number of reactions of a mechanism with reaction intermediates as key species (after Gauss–Jordan reduction)
$K_k^{I(1)}$	Equilibrium constant based on hypothetical ideal gas behaviour at $P = 1$ atm
K_k^0	Equilibrium constant based on pure species at T, P
$K_{y,k}$	Equilibrium constant of the kth canonical reaction in terms of mole fractions
$K_{C,k}$	Equilibrium constant of the kth canonical reaction in terms of molar concentrations (mol m^{-3})$^{\Delta\sigma_k}$
$K_{P,k}$	Equilibrium constant of the kth canonical reaction in terms of partial pressures (atm)$^{\Delta\sigma_k}$
$K_{\phi,k}$	Factor of fugacity coefficients in the equilibrium constant of the kth canonical reaction
$K_{\gamma,k}$	Factor of activity coefficients in the equilibrium constant of the kth canonical reaction
L	Reactor length (axial dispersion model) (m)
ℓ	Slab thickness (m)
M	Non-singular transformation matrix (Chap. 1); mass of the liquid mixture in a tank reactor (kg) (Chap. 2)
M_{GJ}	Transformation matrix associated with the Gauss–Jordan reduction
m_j	Molar mass of A_j (kg mol^{-1})
$\overline{m}$	Mean molar mass of the fluid (kg mol^{-1})

N	Number of components species (Chap. 1); total number of moles (Chap. 2)
N_β	Rank of the atomic matrix
N_j	Number of moles of A_j
N_j^*	Reference number of moles of A_j, Eq. (2.97) in Chap. 2
$\mathcal{N}$	Matrix ($S \times N$), which columns form a basis of the null space of ν
P	Pressure (Pa)
P_j	Partial pressure of A_j (Pa)
$Pe_{j,L} = GL/(\rho D_j)$	Axial Peclet number of A_j in the ADM
$Pe_{m,j} = Gd_p/(\rho \mathfrak{D}_j)$	Molecular Peclet number of A_j
Q^*	Heat transfer rate (W)
q	Volumetric flow rate (m^3 s^{-1})
q_v	Volumetric heat transfer rate (W m^{-3})
$Re = Gd_t/\mu$	Reynolds number
$Re_p = Gd_p/\mu$	Particle Reynolds number
R	Number of a given set of reactions; the ideal gas constant, 8.31446261815324 J mol^{-1} K^{-1}, in Chap. 2
R_{ERS}	Number of elementary steps of a given reaction mechanism
$\mathcal{R}_i$	ith reaction
$\mathfrak{R}_j$	Total production rate of A_j in the entire reactor (mol/s)
r_i	Reaction rate of the ith reaction (mol m^{-3} s^{-1})
$\mathscr{r}_j$	Net molar-production rate of A_j (mol m^{-3} s^{-1})
$\hat{\mathscr{r}}_j$	Net mass-production rate of A_j (kg m^{-3} s^{-1})
S	Number of species in the fluid
$Sc_j = \mu/(\rho \mathfrak{D}_j)$	Schmidt number of species A_j
T	Temperature (K)
$\mathbb{T}$	Component variable associated with the mixture temperature, T (K)
t	Time (s)
$u = G/\rho$	Mass-average velocity (m s^{-1})
V	Reactor volume (m^3)
V_j	Variable related to the mass of A_j
V_j'	Variable related to the moles of A_j
$\mathbb{V}_j$	Component variable associated with V_j
$\mathbb{V}_j'$	Component variable associated with V_j'
W	Mass flow rate (kg s^{-1})
x_A	Conversion of species A
X_i	ith extent of reaction (mol m^{-3})
y_j	Molar fraction of A_j
$\mathbb{Y}_j$	Variable associated with y_j, Eq. (2.13a) in Chap. 2

$\mathcal{y}_j$	Relative variable respect to the inlet molar flow rate, F_j/F_0 (continuous-flow reactors) or to the initial number of moles, N_j/N_I (batch reactors)
$\mathcal{y} = \sum_j \mathcal{y}_j$	Summation of the $\mathcal{y}_j$ values over all species
z_i	Cartesian coordinates (m)
z	Axial coordinate (m)

Greek Letters

α	Coefficient of linear combination between reactions (Eq. 1.26a) in Chap. 1
β	Atomic matrix ($E \times S$)
β_{ej}	Number of the atomic element "e" in species A_j
ΔG_k	Free energy of reaction for the kth canonical reaction, (J mol^{-1})
ΔH_i	Reaction enthalpy of the ith reaction (J mol^{-1})
ΔH_k	Reaction enthalpy of the kth canonical reaction (J mol^{-1})
$\Delta \hat{H}_k$	Specific reaction enthalpy of the kth canonical reaction (J kg^{-1})
ΔH_0^d	Heat of mixing per total mole of the inlet stream (J mol^{-1})
$\Delta \hat{H}_0^d$	Specific heat of mixing of the inlet stream (J kg^{-1})
$\Delta \hat{h}_j^d$	Specific heat of mixing of A_j (J kg^{-1})
ΔT^d	Temperature rise associated with the heat of mixing, Eq. (2.108c) in Chap. 2 (K)
$\Delta\sigma_k = 1 + \sum_{j>K}\sigma_{kj}$	Molar change of the kth normalised canonical reaction
ε	Expansion factor, $\varepsilon = \rho_0/\rho$ (continuous-flow reactor) or $\varepsilon = \rho_I/\rho$ (batch reactor) (Chap. 2); porosity of a packed bed (Chap. 3)
ε_p	Porosity of a porous medium
$\phi_j = y_j f_j/P$	Fugacity coefficient of A_j
φ	Dispersive heat-flux (W m^{-2})
$\gamma_j = y_j f_j/f_j^0$	Activity coefficient of A_j
κ_j	Proportionality coefficient in the definition of V_j (Sect. 3.3 in Chap. 3)
Λ	Effective thermal conductivity (W m^{-2}) (Chap. 3)
Λ_{conv}	Contribution to Λ from flow effects in packed bed (W m^{-2}) (Chap. 3)
$\Lambda_{m,i}$	Coefficients of linear combination between reactions, defined in Eqs. (1.31a) and (1.31b) of Chap. 1
$\lambda_{m,i}$	Coefficients of linear combination between reactions, defined in Eq. (1.31c) of Chap. 1

λ	Thermal conductivity (W m^{-2}) (Chap. 3)
μ	Viscosity (Pa s)
μ_j	Chemical potential of A_j (J mol^{-1})
θ	Residence time or ratio M/W in a semi-batch reactor (s)
ν	Matrix ($R \times S$) of stoichiometric coefficients
ν_{ij}	Stoichiometric coefficient of A_j in the ith reaction
$\hat{\nu}_{ij} = m_j \nu_{ij}$	Mass stoichiometric coefficient of A_j in the ith reaction (kg mol^{-1})
ρ	Density (kg m^{-3})
σ	Matrix ($K \times N$) of stoichiometric coefficients σ_{kj}
σ_{kj}	Stoichiometric coefficient of A_j in the kth canonical reaction
$\hat{\sigma}_{kj} = (m_j/m_k)\sigma_{kj}$	Mass stoichiometric coefficient of A_j in the kth canonical reaction
$\upsilon = 1/C$	Molar volume (m^3 mol^{-1})
υ_j	Partial molar volume of A_j (m^3 mol^{-1})
τ	Tortuosity factor
$\Omega_E = \sum_e \Omega_e$	Equation (3.25a) in Chap. 3
Ω_j	Component variable associated with ω_j
$\Omega_j^* = \Omega_j/(1 - \Omega_E)$	Modified component variable associated with ω_j
ω_j	Mass fraction of A_j
Ψ_j	Component variable associated with ψ_j (mol kg^{-1})
$\psi_j = \omega_j/m_j$	Moles of A_j per unit fluid mass (mol kg^{-1})
ζ	Coordinate across a slab (m)

Subscripts

0	Denotes input-stream value, also value at the interface (Sect. 3.5 in Chap. 3)
A	Key species in a single reaction
α	Index of the inlet/outlet streams
e	Index of chemical elements (Chap. 1); index of the species significantly involved in interfacial transfer (Chap. 3)
I	Initial condition
j	Index of generic species A_j
k, k'	Specific indices of key species A_k
l, m	Indices in the elementary reaction combination, Eq. (1.26a) in Chap. 1
L	Denotes evaluation at the reactor exit (ADM)
ℓ	Denotes evaluation at the slab internal-surface
n	Denotes component normal to a given surface
$Taylor$	Contribution from the effect of velocity profiles to the axial dispersion coefficient

v	Vapour phase
w	Denotes evaluation at a solid boundary

Superscripts

0	Reference state based on pure species at T, P
c	Current value in the elementary reaction combination, Eq. (1.26a) in Chap. 1
$I(1)$	Reference state based on hypothetical ideal gas behaviour at P =1 atm
n	Updated value in the elementary reaction combination, Eq. (1.26a) in Chap. 1
r	Reduced matrix
ref	Generic reference condition for chemical equilibrium
t	Turbulent contribution of the effective transport coefficients

Special Symbols

underline	Vector quantity
overline	Volumetric average
$\cdot$	Scalar product of vectors
∇	Gradient vector

Acronyms

ADM	Axial dispersion model
BR	Batch reactor
CFD	Computational fluid dynamics
CRE	Chemical reaction engineering
CSTR	Continuous-flow stirred tank reactor
GJ	Gauss–Jordan
LI	Linearly independent
ODE	Ordinary differential equations
PF	Plug-flow
PFR	Plug-flow reactor
PRCFDS	Particle-resolved CFD simulation
RI	Reaction intermediates
SBR	Semi-batch reactor
SS	Steady state

A Review of Alternatives for the Use of Stoichiometry in Multiple Reaction Systems

1

1.1 Introduction

The conservation of chemical elements makes it possible to relate the variations in the quantities of the different observable species during a chemical reaction, according to their elementary molecular formula. *Stoichiometry* is the branch of chemistry dedicated to establishing such relationships.

In the study of chemical reactors, the stoichiometry of the reactions taking place is essential, but it is also necessary to know their rates, which is the subject of *chemical kinetics*. For a homogeneous reaction system, the most convenient way to establish the rate of change in the amount of a species A_j due to chemical reactions is r_j: the *net molar production* per unit time and unit volume of an element in the system, small enough so that the state variables can be considered uniform therein. If the reaction is catalysed on a solid surface, an analogous definition is valid, but in terms of the surface area of the solid.

For the study of chemical reactors, it is normally assumed that reactions and their rates are specified. If a single reaction is considered, the production rates r_j of reactants and products are simply related by their stoichiometric coefficients. In introductory courses of chemical reactions engineering (CRE), such relationships allow describing the behaviour of ideal reactors on the basis of the change in the amount of a single species. When a set of multiple reactions is defined, stoichiometry also allows for establishing a number of (independent) linear relationships between the rates r_j, but the corresponding coefficients should be worked out from the stoichiometry of the reaction set. The main subject of this chapter is how to suitably undertake this problem in the context of an introductory CRE course.

An overview of the treatment of chemical reactions in ideal reactors is given in Sect. 1.2 to contextualise the relevance of multiple reactions stoichiometry. Section 1.3

G. F. Barreto and C. D. Luzi, *Applying Multiple-Reaction Stoichiometry to Chemical Reactor Modelling*, Synthesis Lectures on Chemical Engineering and Biochemical Engineering, https://doi.org/10.1007/978-3-031-42375-8_1

reviews procedures from the literature to specify the relationships between the net production rates $\mathcal{r}_j$. Section 1.4 presents the approach considered most appropriate for its use in introductory CRE courses. The possibility of its extension to the analysis of complex reaction mechanisms is also discussed. Finally, Sect. 1.5 summarises the conclusions of the chapter.

1.2 The Treatment of Chemical Reactions in Ideal Reactors. An Overview

To deal with conservation balances, their resolution and consequent analysis of ideal flow reactors (stirred tanks and tubular reactors with plug flow), in introductory CRE courses it is usually considered that the involved reactions are known, as well as their kinetic expressions.

With such considerations, it is usually first studied the case of a single reaction, which generically can be represented in the form $\mathcal{R} = \sum_{j=1}^{S} \nu_j A_j = 0$, where S is, in general, the total number of species present in the mixture, ν_j is the stoichiometric coefficient of species A_j $(j = 1, \ldots, S)$, which is defined positive for products, negative for reactants and null for inert species, including an eventual homogeneous catalyst or inhibitor. The relationships imposed by stoichiometry can be expressed in terms of the net production rate $\mathcal{r}_j$ (mol s^{-1} m^{-3}) of a generic A_j species relative to that of a particular species $j = \mathrm{A}$, chosen as the *key species*:

$$\mathcal{r}_j/\nu_j = \mathcal{r}_\mathrm{A}/\nu_\mathrm{A} \tag{1.1}$$

or

$$\mathcal{r}_j = \sigma_{\mathrm{A}j}\mathcal{r}_\mathrm{A}; \tag{1.2a}$$

$$\sigma_{\mathrm{A}j} = \nu_j/\nu_\mathrm{A} \tag{1.2b}$$

The uniformity of the ratios $\mathcal{r}_j/\nu_j$ (Eq. 1.1) for all species allows defining the reaction rate r from:

$$\mathcal{r}_j = \nu_j r \tag{1.3}$$

Assuming that the dependence of r on the state variables (composition, temperature and pressure) is known, the expressions (1.1)–(1.3) complete the stoichiometrically imposed relationships to analyse the reaction processed in a given reactor. Taking as an example a plug-flow reactor in steady state (SS), the conservation equations for all species form a set of ordinary differential equations (ODE system):

$$dF_j/dV = \mathcal{r}_j \tag{1.4}$$

where F_j is the molar flow rate of A_j and V is the incremental reactor volume from the feed inlet section.

Combining Eqs. (1.2a) and (1.4), it is obtained $dF_j = \sigma_{Aj} dF_A$, which upon integration from the inlet to a generic section yields:

$$F_j - F_{j0} = \sigma_{Aj}(F_A - F_{A0}). \tag{1.5}$$

Equation (1.5) stoichiometrically relates all molar flow rates variations to that of A, so it is only necessary to keep the conservation equation of this species:

$$dF_A/dV = \mathcal{r}_A \tag{1.6}$$

The ODE system has been reduced to a single differential equation, in terms of a single dependent variable, F_A. Once $F_A(V)$ is obtained, the stoichiometric relations (1.5) allow evaluating the mixture composition all along the reactor. Apart from simplifying the solution procedure, it is relevant to emphasise the conceptual significance of Eq. (1.5): a single variable (F_A) allows determining the evolution of the whole composition of the reacting mixture when the feed composition (F_{j0}) is specified.

If the mixture density remains invariable, $C_j - C_{j0} = \sigma_{Aj}(C_A - C_{A0})$ is obtained from dividing Eq. (1.5) by the volumetric flow rate q. In this particular case, the molar concentration C_A is a suitable variable to define the state of the system, while Eq. (1.6) is rewritten as $q\, dC_A/dV = \mathcal{r}_A$.

A second step in the study of ideal chemical reactors is the treatment of multiple reactions. A set of R reactions can be presented in the form:

$$\mathcal{R}_i = \sum_{j=1}^{S} \nu_{ij} A_j = 0, \qquad i = 1, \ldots, R \tag{1.7}$$

where ν_{ij} is the stoichiometry coefficient of A_j in the ith reaction. For a specific case, the species will be chemically defined, as well as the coefficients ν_{ij} and the kinetic expressions for each reaction rate r_i.

The net production rate of each species is written by adding the contributions of all reactions (1.7):

$$\mathcal{r}_j = \sum_{i=1}^{R} \nu_{ij} r_i, \qquad j = 1, \ldots, S \tag{1.8}$$

For a plug-flow reactor, the conservation equations are formally Eq. (1.4).

In most CRE courses and textbooks, basic concepts of multiple reaction systems (e.g., selectivity, yield, the occurrence of maxima and minima in concentrations) are illustrated through examples with a couple of reactions, either in series or in parallel, and considering simple kinetics expressions, so that Eq. (1.4) can be analytically solved. The simplicity of such examples does not require a formalism to relate the production rates r_j (see, e.g., [1–9]).

For the treatment of more complex reaction sets, for example with $R > 2$, in some textbooks (e.g., [10]) it is emphasised that the usual occurrence of non-linear kinetics requires the resolution of the mass conservation equations by means of a numerical procedure. Then, the suggested approach is the simultaneous solution of the ODE system of Eq. (1.4) for all species on a suitable numerical platform.

While such an approach is clearly feasible, it excludes answering a conceptual question prompted by the single reaction case:

> Since the stoichiometry prescribes that only one variable (e.g., F_A) suffices to determine the composition of the reacting mixture and the species distribution along a plug-flow reactor for a single reaction, does stoichiometry also prescribe a minimum number of variables and balance equations accomplishing similar features for a given set of reactions?

The answer can be found by recognising that, once specifying the reactions (1.7), there are relationships between the net production rates in the form:

$$r_j = \sum_{k=1}^{K} \sigma_{kj} r_k; \qquad j = K+1, \ldots, S \tag{1.9}$$

where K is the number of linearly independent (LI) reactions of the set (1.7), with $K \leq R$, r_k is the net production rate for each A_k species within this set (index $k = 1, \ldots, K$), called *key species*, and σ_{kj} are stoichiometric coefficients derived from the ν_{ij} coefficients of the original system (1.7). Equation (1.9) states that the net rates r_j of the remaining $S-K$ species A_j (index $j = K+1, \ldots, S$) are determined when all the r_k are specified. For a single reaction, $K = 1$, Eq. (1.9) reduces to Eq. (1.2a) and the coefficients σ_{kj} to those defined in Eq. (1.2b), with $k \equiv A$. It should be mentioned that in the literature, the key species ($k = 1, \ldots, K$) are alternatively referred to as *reference, pivot* or *secondary* species, while the remaining ones ($j = K+1, \ldots, S$) are referred to as *non-key, component* or *primary* species. Here, we will use the terms *key species* for the first group and *component species* for the second.

In a general case, determining the number K, the coefficients σ_{kj} and a feasible set of K key species requires a systematic approach based on linear algebra. Assuming that for a given set of reactions as given by Eq. (1.7) the relationships (1.9) have been evaluated, the conservation Eq. (1.4) for a plug-flow reactor can be combined to obtain the following relationships between the molar flow rates:

$$F_j - F_{j0} = \sum_{k=1}^{K} \sigma_{kj}(F_k - F_{k0}); \qquad j = K+1, \ldots, S \tag{1.10}$$

Equation (1.10) replaces the conservation equations for each component species and the ODE system is reduced to the conservation equations for the key species:

$$dF_k/dV = r_k; \qquad k = 1, \ldots, K \tag{1.11}$$

It can be recognised that Eqs. (1.10) and (1.11) are the extension of Eqs. (1.5) and (1.6) for a single reaction. It is also evident that the number K sets the affirmative answer to the previously raised question.

It should be emphasised that, for a given reaction set (1.7), the relationships (1.9) are extensions of the laws of stoichiometry for a single reaction. In this sense, Eq. (1.9) will be valid for any set of reactions, regardless of their complexity and of whether they are volumetric or superficial.

Another alternative to answer the previously raised question is appealing to the *extent of reactions*, as a set of auxiliary variables to represent the system composition (see, e.g., [11–20]). For the plug-flow reactor, the extent of reactions can be defined in units of molar flow rate, X_i, by means of $F_j - F_{j0} = \sum_{i=1}^{R} \nu_{ij} X_i$, $j = 1, \ldots, S$. Replacing the r_j from Eqs. (1.8) to (1.4) and taking into account the definition of X_i, $dX_i/dV = r_i$ is obtained as the governing equation for X_i. The number of auxiliary variables X_i is always equal to R. If the reaction set (1.7) is linearly independent (LI), it will be $K = R$, so the use of extent of reactions will provide the correct answer to the previously raised question.

However, there are many examples where the proposed set of reactions is not LI ($R > K$). An important reason for such an option is the possibility to express the rate r_i of each reaction with the corresponding thermodynamic driving force, which facilitates the interpretation of the kinetics of the system. For example, in a study of the hydrogenation and isomerisation of unsaturated C4 compounds [21], $S = 6$ species are involved in $R = 9$ kinetically characterised reactions. The number of LI reactions is $K = 4$. In the example, the direct use of $R = 9$ extent of reactions not only obscures the fact that just $K = 4$ variables determine the system composition, but also leads to a computationally inefficient solution of $R = 9$ equations, $dX_i/dV = r_i$, even compared to the direct solution of the $S = 6$ Eq. (1.4) for all species. As discussed in some textbooks [11, 12, 16], the use of extent of reactions should be conveniently reduced to the minimum number K. In general, this involves an algebraic analysis of the system (1.7) to determine a subsystem of a number K of LI reactions and "effective" reaction rates for them, which in turn requires expressing the dependent reactions as linear combinations of the chosen LI subset (see, for example, [16]). It should be noted that if $K < R$, such a reduction can be essential in some situations, as for example in the calculation of chemical equilibrium, for which there will be infinite solutions in terms of the extent of reactions of the original reaction set.

It follows that, in order to achieve a generalised description of the reduction of a reaction set, either by the identification of key species or by the use of extent of reactions, it is necessary to appeal to the use of mathematical concepts of linear algebra.

Upon considering the significance of the relationships (1.9) for multiple reactions, which extend the elementary stoichiometric relationships (Eq. 1.2a) for single reactions, the remaining of this chapter explores some literature alternatives for specifying those relationships, as well as a detailed description of our proposed choice. In Chaps. 2 and 3, the concept of key species will be further used in the formulation of the mass conservation equations for the case of multiple reactions and in discussing strategies for their resolution.

1.3 Literature Procedures for Evaluating Production-Rates Relationships

Different alternatives for evaluating the number K of linearly independent reactions and the coefficients σ_{kj} of the stoichiometric relationships (1.9), from a given set of reactions defined according to Eq. (1.7), will be reviewed in Sects. 1.3.1–1.3.3.

1.3.1 Procedure I

The following procedure to evaluate the coefficients σ_{kj} in Eq. (1.9) can be found in some textbooks [11–13, 22, 23]. The matrix $\nu(R \times S) = [\nu_{ij}]$ of stoichiometric coefficients is defined with rows corresponding to reactions and columns to species. Then, Eq. (1.8) is written as follows:

$$\underline{\mathscr{r}} = \nu^T \underline{r} \tag{1.12}$$

where $\underline{\mathscr{r}}$ is the column vector of the S net production rates $\mathscr{r}_j$, and $\underline{r}$ is the column vector of the R reaction rates, r_i. The rows of the matrix $\nu^T(S \times R)$ then correspond to each of the S species and the columns to each of the R reactions.

It is assumed that in a given reacting system, the net velocities $\mathscr{r}_j$ have been **accurately** measured and the r_i values are to be evaluated. Linear algebra tells us that depending on the rank K of the matrix ν^T (i.e., the rank of ν), which is always smaller than S, the following situations can be found. The system (1.12) will have a unique solution if $K = R$, in which case ν^T will have a left inverse. On the contrary, if $K < R$ the system will have infinite solutions (we know, by construction, that the system is compatible) and there will be $R - K$ degrees of freedom to determine the remaining K unknowns r_i. Therefore, in a first step, the rank K of ν^T must be evaluated. It is assumed, on the one hand, that the ordering of the columns of ν^T is chosen such that the first K columns are linearly independent. This implies that the corresponding reactions are linearly independent (LI), while the remaining $(R - K)$ reactions are linear combinations of the first K reactions. For the purpose of solving the system (1.12) under the above conditions, the last $(R - K)$ columns of ν^T and their corresponding rates r_i of the vector $\underline{r}$ must be eliminated. The system (1.12) is then restated in the form:

$$\underline{\mathcal{r}} = (\nu_{K,S})^T \underline{r}_K \tag{1.13}$$

where the matrix $(\nu_{K,S})^T$ only maintains the first K columns of ν^T and the elements of $\underline{r}_K$ are the corresponding K unknowns r_i. Additionally, without loss of generality, it is assumed that the rows of $(\nu_{K,S})^T$ are ordered such that the first K rows are LI. The matrix $(\nu_{K,S})^T$ can be partitioned in the form:

$$(\nu_{K,S})^T = \begin{bmatrix} (\nu_{K,K})^T \\ (\nu_{K,N})^T \end{bmatrix} \tag{1.14}$$

where $N = S - K$. With such a partition, the system (1.13) is now written as:

$$\underline{\mathcal{r}}_K = (\nu_{K,K})^T \underline{r}_K \tag{1.15a}$$

$$\underline{\mathcal{r}}_C = (\nu_{K,N})^T \underline{r}_K \tag{1.15b}$$

where $\underline{\mathcal{r}}_K$ is the vector of net production rates of the first K species, ordered according to Eq. (1.15a), which will correspond to the key species, and $\underline{\mathcal{r}}_C$ is the vector of net production rates of the last N species, identified as component species.

Since for the chosen ordering $(\nu_{K,K})^T$ is non-singular, Eq. (1.15a) allows to write $\underline{r}_K = (\nu_{K,K}^{-1})^T \underline{\mathcal{r}}_K$, which is replaced in (1.15b):

$$\underline{\mathcal{r}}_C = (\nu_{K,K}^{-1} \nu_{K,N})^T \underline{\mathcal{r}}_K \tag{1.16}$$

Equation (1.16) is equivalent to the set of scalar relationships (1.9). Then, defining the matrix $\sigma(K \times N)$ of coefficients σ_{kj}, $\sigma = [\sigma_{kj}]$, it follows from Eq. (1.16) that $\sigma^T = (\nu_{K,K}^{-1} \nu_{K,N})^T$.

Different combinations can be chosen for the sets of key and component species, which will be determined by all possible combinations that allow the first rows of $(\nu_{K,S})^T$ to be linearly independent.

Although the procedure proposes a hypothetical situation in which the elements of $\underline{\mathcal{r}}$ are given and those of $\underline{r}$ are unknowns, the evaluation of the coefficients σ_{kj} in the relationships (1.9) is obtained as a result. A few points are worth commenting on:

- If the net velocities $\mathcal{r}_k$ are actually measured, it will only be possible to evaluate K values r_i of the proposed system of R reactions, *i.e.*, the number of LI reactions. Considering that the system evolves according to the proposed R reactions, the r_i values **are not necessarily** the actual values of the chosen subset of LI reactions but represent "effective values" that allow the $\mathcal{r}_k$ measurements to be satisfied.
- If the known reaction-rate expressions r_i of a given set of R reaction are used to study the behaviour of a given reactor, it will be necessary to evaluate the elements of the vector $\underline{\mathcal{r}}_K$ using the values of **all** reaction rates r_i in Eq. (1.8).

– The steps described to reach Eq. (1.16) can be easily assimilated by those who are familiar with elements of linear algebra and reaction systems. However, such a condition will not be guaranteed for most students in an introductory CRE course. The above steps should be exemplified and concepts such as linear independence of reactions, the connection with the rank of the matrix ν^T and the ordering of the matrix ν^T in such a way that the submatrix $(\nu_{K,K})^T$ is non-singular should be recalled. On the other hand, from a practical point of view, the determination of the inverse of $(\nu_{K,K})^T$ and the product $\nu_{K,K}^{-1}\nu_{K,N}$ will require the use of a calculus platform, even in simple cases.

1.3.2 Procedure II

The second procedure has been employed and analysed by Björnbom [24] and is also presented in Chap. 2 of the book of Smith and Missen [25].

First, it is considered a matrix $\mathcal{N}(S \times N)$, which columns form a basis of the null space of the stoichiometric matrix ν, i.e., satisfying $\nu\mathcal{N} = 0$, and $N = S - K$ is simultaneously the rank of $\mathcal{N}$ (since the rank of ν is K) and the dimension of the null space of ν.

Multiplying Eq. (1.12) by $\mathcal{N}^T$ gives $\mathcal{N}^T \underline{\mathscr{r}} = \mathcal{N}^T \nu^T \underline{r} = (\nu\mathcal{N})^T \underline{r}$, from which it follows that:

$$\mathcal{N}^T \underline{\mathscr{r}} = \underline{0} \tag{1.17}$$

Any procedure to evaluate the coefficients σ_{kj} in the relationships (1.9) can be expressed in the form of Eq. (1.17). In the case of Procedure I, although this property was not explicitly employed, it can be identified from Eq. (1.16) that the matrix $\mathcal{N}$ involved is:

$$\mathcal{N} = \begin{bmatrix} \nu_{K,K}^{-1}\nu_{K,N} \\ \cdots\cdots\cdots\cdots \\ -I_N \end{bmatrix},$$

where I_N is the identity matrix of $N \times N$ elements.

Instead, Procedure II makes explicit use of the evaluation of a convenient matrix $\mathcal{N}$. For this purpose, the matrix ν^r resulting from the Gauss–Jordan (GJ) reduction of the matrix ν is evaluated, which can always be expressed as in the following configuration:

$$\nu^r = \begin{bmatrix} I_K & : & \sigma \\ \cdots\cdots\cdots & : & \cdots\cdots\cdots \\ 0_{R-K,K} & : & 0_{R-K,N} \end{bmatrix}, \tag{1.18}$$

where I_K is the identity matrix of $K \times K$ elements and the resulting submatrix $\sigma(K \times N)$ is just the matrix $\sigma = [\sigma_{kj}]$ of coefficients σ_{kj} in Eq. (1.9), as will be confirmed below.

The submatrices $0_{R-K,K}$ and $0_{R-K,N}$ are null matrices arising when $R - K$ reactions are linearly dependent on the remaining K reactions. Recalling that the null space of ν^r is the same as that of ν, it arises from (1.18) that the columns of

$$\mathcal{N} = \begin{bmatrix} \sigma \\ \cdots\cdots \\ -I_N \end{bmatrix} \tag{1.19}$$

correspond to a basis of the null space of ν, where I_N is the identity matrix of $N \times N$ elements. Using $\mathcal{N}$ from Eqs. (1.19) to (1.17), and partitioning the vector $\underline{\mathscr{r}}$ as:

$$\left[\sigma^T \;\vdots\; -I_N\right] \begin{bmatrix} \underline{\mathscr{r}}_K \\ \cdots\cdots \\ \underline{\mathscr{r}}_C \end{bmatrix} = \underline{0} \tag{1.20}$$

It follows that

$$\underline{\mathscr{r}}_C = \sigma^T \underline{\mathscr{r}}_K \tag{1.21}$$

Equation (1.21) is equivalent to the set of Eq. (1.9) with coefficients σ_{kj} given by the elements of the submatrix σ in the partition of ν^r (Eq. 1.18).

It is worth noting a property of the coefficients σ_{kj}. Consider the product $(M\nu) = \nu^n$, where $M(R \times R)$ is a non-singular matrix. The rows of the matrix ν^n are linear combinations of the rows of ν and retain their dimension K. Therefore, the rows of ν^n constitute a new set of K linearly independent reactions obtained from the original set of reactions defined by the rows of ν. If the new set of reactions is to be used to represent the behaviour of the system, the net generation rates of the species $\mathscr{r}_j$ must remain unchanged. This condition requires defining appropriate values for the reaction rates of the new set of reactions, which is always possible, as presented below. For this purpose, Eq. (1.12) is rewritten in the form:

$$\underline{\mathscr{r}} = \left(M^{-1}M\nu\right)^T \underline{r} = (M\nu)^T \left(M^{-1}\right)^T \underline{r} = \left(\nu^n\right)^T \underline{r}^n \tag{1.22}$$

Therefore, the elements r_i^n of the vector $\underline{r}^n$ results in $\underline{r}^n = \left(M^{-1}\right)^T \underline{r}$.

For the GJ reduction leading to the reduced matrix ν^r in Eq. (2.18), the specific transformation matrix M is denoted as M_{GJ}. Then, the GJ reduction procedure can be formulated as $M_{GJ}\nu = \nu^r$, from which it follows that the rows of the upper submatrix $[I_K \vdots \sigma]$ in (2.18) represent a new set of K linearly independent reactions. The K species A_k, $k = 1, \ldots, K$, associated with the columns of I_K are the resulting key species and each of them only participates in a single reaction (that of the kth row) with stoichiometric coefficient equal to one. Such a set of reactions is called a set of *canonical reactions*. It should be emphasised that the distinguishing feature of a canonical reaction set is that there is a biunivocal correspondence between reactions and key species, while the unit stoichiometric coefficient of each key species is a convenient standardisation. It is also noted that the rate of each canonical reaction $\mathcal{R}_k$ is $\mathscr{r}_k$, which must be evaluated from all the original reaction rates r_i, using (1.8) for each key species A_k.

Although Procedure II involves an exhaustive stoichiometric analysis, it can be concluded that it not only solves the problem of determining the coefficients σ_{kj}, but also provides additional valuable information. On the one hand, it simultaneously introduces the concept of linear independence of reactions, in particular with respect to the originally proposed set. In turn, it follows that it is possible to associate the coefficients σ_{kj} with the stoichiometric coefficients of a set of canonical reactions, capable of representing the behaviour of the reacting system as an alternative to the original reaction system.

However, with respect to teaching practice, it is evident that the present procedure employs an important variety of linear algebra concepts, a fact that makes its adoption difficult, at least in introductory CRE courses. In this sense, it is worth mentioning that Procedure II is not included, to our knowledge, in any undergraduate CRE textbook.

1.3.3 Procedure III

As will be seen below, Procedure III is not entirely equivalent to the previous ones. To this end, it is necessary to recall that from the beginning of this chapter it is proposed to analyse reacting systems with a set of specified reactions. A different scenario arises when a new reacting system is studied, in preliminary instances where the reacting species have been identified and the maximum number of LI reactions involving such species is attempted to be determined. In such a case, the procedure discussed in this section allows determining the stoichiometric relationships between the net production rates associated with the LI reactions found [18, 25, 26]. However, we will see that the number of stoichiometric relationships may not coincide with that resulting from the problem in which the set of reactions is defined beforehand, as is the case for Procedures I and II previously described.

Procedure III is based on the conservation of chemical elements in a chemical reaction. Then, let E be the total number of chemical elements present in the identified S reacting species. If electrically charged species are included, E must incorporate the electric charge as an additional "element", with positive values for protons and negative values for anions. Let β_{ej} be the number of a given element e in the species A_j, and ν_j the stoichiometric coefficients of the species A_j in a given reaction. Then, the conservation of elements is formulated as $\sum_{j=1}^{S}(\beta_{ej}\nu_j) = 0$ for $1 \le e \le E$. By arranging the coefficients ν_j in the row vector $\underline{\nu}$ of S components, and defining the *atomic matrix* $\beta(E \times S) = [\beta_{ej}]$, the above conservation equations can be written as:

$$\beta \underline{\nu}^T = \underline{0} \tag{1.23a}$$

According to Eq. (1.23a), $\underline{\nu}^T$ belongs to the null space of the matrix β. Then, having identified a basis of the null space of β (e.g., from a GJ reduction), it can be arranged in a matrix ν^T of $S \times (S - N_\beta)$ elements, where N_β is the rank of the matrix β. Thus, the columns of ν^T define the stoichiometric coefficients of each LI reaction.

Additionally, the net production rates of the S species r_j also verify the conservation of each of the elements. That is, $\sum_{j=1}^{S}(\beta_{ej} r_j) = 0$ for $1 \leq e \leq E$, a fact that can be expressed by:

$$\beta \underline{r} = \underline{0} \tag{1.23b}$$

The vector $\underline{r}$ satisfying (1.23b) clearly belongs to the null space of β, whose dimension is $S \times (S - N_\beta)$ as noted above. Therefore, the system (1.23b) allows expressing N_β elements of the vectors $\underline{r}$ in terms of the remaining $(S - N_\beta)$ elements, i.e., N_β relationships of the type (1.9) between the net generation rates r_j will be determined. The same GJ reduction as in Eq. (1.23a) will accomplish the purpose of specifying N_β independent relationships of type (1.9).

However, the number $(S - N_\beta)$ of LI reactions satisfying (1.23a) may be larger than the number K of LI reactions that can actually take place in the system, for example, because of kinetic constraints. Thus, for a system involving S reactive species, it will be verified in general that $K \leq (S - N_\beta)$. Consequently, the number of relationships (1.9) determined in the previous sections as $N = (S - K)$, may be higher than the number N_β arisen from Eq. (1.23b), $N \geq N_\beta$.

Practical examples with $N > N_\beta$ are given, e.g., by Björnbom [27]. A typical case is the hydro-dealkylation of toluene (T) to produce benzene (B) and methane (M). This single reaction ($K = 1$) involves $S = 4$ species (H_2, T, B, M) and $E = 2$ elements (C, H). Therefore, $N_\beta = 2$, but $N = S - K = 3$. A second independent reaction, which is not observed in the actual practice, can be the total hydrogenation of toluene to give methane.

1.4 Proposed Strategy for Specifying the Relationships Between Production Rates

As pointed out in the previous section, procedures such as I and II for specifying relations of the type (1.9) involve a variety of linear algebra concepts and manipulations, which will be difficult to adopt by the teaching practice of an introductory CRE course. Hence, we present in this section a strategy to achieve the same results, which in the authors' experience is more natural and effective for implementation in an introductory CRE course.

On the one hand, the specification of the relationships (1.9) is developed on the basis of a suitable specific example and the procedure is generalised from it. On the other

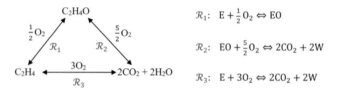

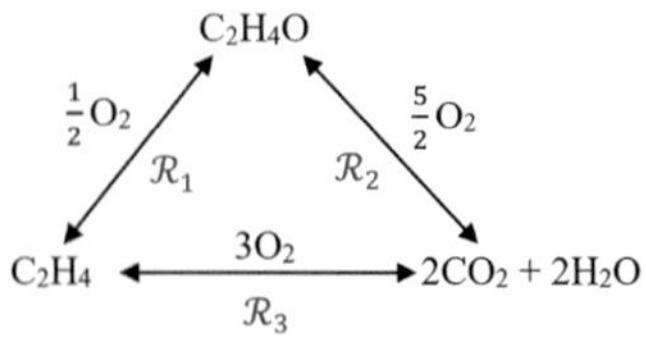

$$\mathcal{R}_1:\ E + \tfrac{1}{2}O_2 \Leftrightarrow EO$$

$$\mathcal{R}_2:\ EO + \tfrac{5}{2}O_2 \Leftrightarrow 2CO_2 + 2W$$

$$\mathcal{R}_3:\ E + 3O_2 \Leftrightarrow 2CO_2 + 2W$$

Fig. 1.1 Reactions involved in the production of ethylene oxide [28]

hand, the example appeals to the concept of linear combination of reactions to reach an alternative system of canonical reactions (defined in Sect. 1.3.2). Such combinations are performed following the same steps as those involved in the GJ reduction of the stoichiometric matrix ν, which allows the conclusions to be generalised by recreating only the concept and significance of the rank of a matrix. In this sense, it will be seen that the proposed strategy does not require the prior concept of linear independence of reactions, but that it will be an outcome from the procedure itself. In the aforementioned terms, the proposal will be presented here with a degree of detail advisable as a reference guide for students. To conclude this introductory paragraph, it is anticipated that Sects. 1.4.2 and 1.4.3 present extensions to the given procedure for covering additional aspects to the one involved in the specification of the relationships (1.9).

As a control case to be analysed, we consider the case of $R = 3$ reactions proposed by Rebsdat and Mayer [28] to describe the production of ethylene oxide (EO) from the partial oxidation of ethylene (E) accompanied by the undesirable reactions of complete combustion of E and EO, as schematised in Fig. 1.1, where W represents H_2O.

Rebsdat and Mayer [28] also provide a brief summary of the kinetic behaviour of all three reactions and give additional references of studies discussing kinetic data and models. Here, we simply assume that the form and parameters of each of the r_i are available. The net production rates $\mathscr{r}_j$ (Eq. 1.8) for the $S = 5$ species are expressed by:

$$\begin{aligned}
\mathscr{r}_{E} &= -r_1 - r_3 \\
\mathscr{r}_{EO} &= r_1 - r_2 \\
\mathscr{r}_{CO_2} &= 2r_2 + 2r_3 \\
\mathscr{r}_{W} &= 2r_2 + 2r_3 \\
\mathscr{r}_{O_2} &= -\frac{1}{2}r_1 - \frac{5}{2}r_2 - 3r_3
\end{aligned} \tag{1.24}$$

First, it should be recalled the concept of a canonical reaction system, defined when in each reaction one species can be identified that does not participate in the others. Furthermore, a canonical system is *normalised* when the stoichiometric coefficients of such species are changed to 1 (one) (Sect. 1.3.2). It is immediate that the example in Fig. 1.1 does not satisfy the definition. Any of the species participates in more than one reaction, a fact that also arises from Eq. (1.24): any of the $\mathscr{r}_j$ depends on at least two r_i values. However, if for a moment it is assumed that reaction $\mathcal{R}_3$ does not occur (i.e.,

$r_3 = 0$), it follows that $\mathcal{R}_1$ and $\mathcal{R}_2$ will be canonical with respect to, e.g., E and CO_2, respectively. The system of Eq. (1.24) reduces to:

$$\mathcal{r}_{E} = -r_1; \qquad \mathcal{r}_{CO_2} = 2r_2$$

$$\mathcal{r}_{EO} = r_1 - r_2; \qquad \mathcal{r}_{W} = 2r_2; \qquad \mathcal{r}_{O_2} = -\frac{1}{2}r_1 - \frac{5}{2}r_2$$

As proportionality holds between $\mathcal{r}_E$ and r_1, and between $\mathcal{r}_{CO_2}$ and r_2, the values of r_1 and r_2 can be replaced in the remaining expressions to obtain:

$$\begin{aligned} \mathcal{r}_{EO} &= -\mathcal{r}_{E} - \frac{1}{2}\mathcal{r}_{CO_2} \\ \mathcal{r}_{W} &= \mathcal{r}_{CO_2} \\ \mathcal{r}_{O_2} &= \frac{1}{2}\mathcal{r}_{E} - \frac{5}{4}\mathcal{r}_{CO_2} \end{aligned} \tag{1.25}$$

Equation (1.25) express relationships between the $\mathcal{r}_j$ in the desired general form of Eq. (1.9). It can also be easily concluded that it is not possible to establish additional relationships between the $\mathcal{r}_j$ values. Since the r_1 and r_2 values must be independent of each other, $\mathcal{r}_E$ and $\mathcal{r}_{CO_2}$ will also be independent of each other. The coefficients σ_{kj} are specified, with EO, W and O_2 as component species, E and CO_2 as the key species, and $K = 2$ is the number of canonical reactions, coincident with that of key species. Furthermore, this procedure is easily generalised for any system of R canonical reactions, with $K = R$, the key species being those associated with each canonical reaction and in Eq. (1.9) the coefficients evaluated as $\sigma_{kj} = \nu_{kj}/\nu_{kk}$.

Returning to the original example ($r_3 \neq 0$), when the system of reactions is not canonical, one can proceed with a sequence of linear combinations between the reactions, so as to transform the original system into an equivalent system of canonical reactions. At each step of the sequence, a given reaction $\mathcal{R}_m^c$ is replaced by a new one, $\mathcal{R}_m^n$, resulting from the combination of $\mathcal{R}_m^c$ with some other reaction of the system $\mathcal{R}_l^c$, in the form:

$$\mathcal{R}_m^n = \alpha_m \mathcal{R}_m^c + \alpha_l \mathcal{R}_l^c \tag{1.26a}$$

where α_m and α_l are coefficients chosen according to convenience, with the only restriction that $\alpha_m \neq 0$. Once $\mathcal{R}_m^n$ is defined according to (1.26a), the new set retains the rest of the reactions and $\mathcal{R}_m^n \rightarrow \mathcal{R}_m^c$ is reassigned. At the beginning, the set $\{\mathcal{R}_i^c\}$ coincides with the original system of reactions $\{\mathcal{R}_i\}$, but as the operations (1.26a) proceed, the system is updated to a new set $\{\mathcal{R}_m^c\}$ identified by the superscript "c". A transformation of the type (1.26a) is applied in practice on the stoichiometric coefficients of each species:

$$\nu_{mj}^n = \alpha_m \nu_{mj}^c + \alpha_l \nu_{lj}^c; \qquad j = 1, \ldots, S \tag{1.26b}$$

It was seen in Sect. 1.3.2 that a non-singular transformation of a given reaction set leads to a new set that preserves the values of generation rates $\mathcal{r}_j$. However, a specific

$$\nu:\ \begin{array}{c} \\ \mathbf{1} \\ \mathbf{2} \\ \mathbf{3} \end{array}\begin{array}{c} \begin{array}{ccccc} EO & CO_2 & E & W & O_2 \end{array} \\ \begin{bmatrix} 1 & 0 & -1 & 0 & -\frac{1}{2} \\ -1 & 2 & 0 & 2 & -\frac{5}{2} \\ 0 & 2 & -1 & 2 & -3 \end{bmatrix} \end{array} \xrightarrow[\text{reduction}]{\text{After GJ}} \nu^r:\ \begin{array}{c} \\ \mathbf{1} \\ \mathbf{2} \\ \mathbf{3} \end{array}\begin{array}{c} \begin{array}{ccccc} EO & CO_2 & E & W & O_2 \end{array} \\ \begin{bmatrix} 1 & 0 & -1 & 0 & -\frac{1}{2} \\ 0 & 1 & -\frac{1}{2} & 1 & -\frac{3}{2} \\ 0 & 0 & 0 & 0 & 0 \end{bmatrix} \end{array}$$

Canonical reactions

$$\mathcal{R}_{EO} = \mathbf{EO} - 1E + 0W - \frac{1}{2}O_2 = 0, \quad \text{or} \quad E + \frac{1}{2}O_2 \Leftrightarrow \mathbf{EO} \tag{F1.a}$$

$$\mathcal{R}_{CO_2} = \mathbf{CO_2} - \frac{1}{2}E + 1W - \frac{3}{2}O_2 = 0, \quad \text{or} \quad \frac{1}{2}E + \frac{3}{2}O_2 \Leftrightarrow \mathbf{CO_2} + W \tag{F1.b}$$

Relationships between net generation rates

$$r_E = -1r_{EO} - \frac{1}{2}r_{CO_2}; \quad r_W = 0r_{EO} + 1r_{CO_2}; \quad r_{O_2} = -\frac{1}{2}r_{EO} - \frac{3}{2}r_{CO_2}; \tag{F2}$$

Fig. 1.2 Example of GJ reduction for the stoichiometric matrix ν, corresponding to reactions in Fig. 1.1

alternative to arrive at the same conclusion from the combination (1.26a), but avoiding the use of matrix operations, is put forward in Appendix 1. Therefore, the relationships in the form of Eq. (1.9) finally obtained on the basis of the reaction set already transformed into canonical reactions, will also be valid for the original system $\{\mathcal{R}_i\}$. It is worth mentioning that in Appendix 1 the restriction $\alpha_m \neq 0$ in Eqs. (1.26a, b) is also validated, although it can be intuitively accepted by the need to maintain the incidence of reaction $\mathcal{R}_m^c$ on the system.

To systematise the sequence of operations (1.26a, b), the original coefficients ν_{ij} are arranged in the rows of the stoichiometric matrix ν. In principle, any arrangement of reactions (rows) and species (columns) is valid, but the chosen arrangement must be recorded. For the system in Fig. 1.1, a specific configuration is presented in the upper left part of Fig. 1.2.

The combinations (1.26a, b) correspond to combinations between the rows of ν, with the purpose of removing the participation of a given species A_j from a given reaction $\mathcal{R}_m^c$, by appropriate choice of the reaction $\mathcal{R}_l^c$ and of the α_m and α_l coefficients. In the example, and with the chosen arrangement of the matrix ν, the final purpose is to obtain a transformed matrix with the following configuration:

$$\begin{array}{c} \\ \mathbf{1} \\ \mathbf{2} \\ \mathbf{3} \end{array}\begin{array}{c} \begin{array}{ccccc} EO & CO_2 & E & W & O_2 \end{array} \\ \left[\begin{array}{ccc:cc} 1 & 0 & 0 & \sigma_{EO,W} & \sigma_{EO,O_2} \\ 0 & 1 & 0 & \sigma_{CO_2,W} & \sigma_{CO_2,O_2} \\ 0 & 0 & 1 & \sigma_{E,W} & \sigma_{E,O_2} \end{array}\right] \end{array} \tag{1.27}$$

If such a configuration is feasible, each of the species EO, CO_2 and E will participate in only one reaction with stoichiometric coefficient "1", so that the transformed reaction set will be canonical and normalised with EO, CO_2 and E as key species, while the coefficients σ_{kj} for the resulting component species, W and O_2 in the present case, will be defined by the elements of the respective columns (it must be remembered that to obtain a canonical set of reactions it will only be necessary to reach non-zero coefficients on the main diagonal).

If the configuration (1.27) is reached, it must be recognised that it corresponds to the result of the GJ reduction of ν. Therefore, the GJ reduction algorithm can be invoked and identify each stage with the sequence of operations (1.26a, b). Thus, we proceed by reducing the columns from the left. For each column j, the element ν^c_{jj}, assumed to be non-zero, is called the pivot. It is first transformed into a "1" and then the remaining elements of column j are reduced to zero. In the first column of ν (Fig. 1.2), the coefficient $\nu_{1,1}$ is fortuitously "1". Otherwise, it would have been necessary to apply (1.26a, b) to replace the first row by $\mathcal{R}^n_1 = \mathcal{R}^c_1/\nu^c_{1,1}$ (i.e., $\nu^n_{1,j} = \nu^c_{1,j}/\nu^c_{1,1}$; $j = 1, \ldots, S$) (the treatment if $\nu_{1,1} = 0$ is recalled later). Successive replacements of each of the remaining rows ($i \neq 1$) by combinations with the first one, applying (1.26a, b) in the form $\mathcal{R}^n_i = \mathcal{R}^c_i - \nu^c_{i,1}\mathcal{R}^c_1$ (i.e., $\nu^n_{i,j} = \nu^c_{i,j} - \nu^c_{i,1}\nu^c_{1,j}$; $j = 1, \ldots, S$), allow the transformation of the first column to be completed. The reduction should proceed similarly with column 2 and 3. However, after reducing the second column, the result indicated by the matrix ν^r in the upper right of Fig. 1.2 is reached, where all the elements of the third row have become zero. This result is a consequence of the fact that the original reaction $\mathcal{R}_3$ (identified by the label of the third row) is a linear combination of the original reactions $\mathcal{R}_1$ and $\mathcal{R}_2$. By inspection (Fig. 1.2), it is easily concluded that $\mathcal{R}_3 = \mathcal{R}_1 + \mathcal{R}_2$.

At this stage, it should be recalled that the number of "1's" on the main diagonal after the GJ reduction corresponds to the rank K of the matrix. In the example, $K = 2$ indicates that ν has two LI rows, corresponding to the original reactions $\mathcal{R}_1$ and $\mathcal{R}_2$ (identified by their labels), which then are LI to each other. Furthermore, the elements of the first two rows of ν^r correspond to the stoichiometric coefficients of the $K = 2$ canonical reactions $\mathcal{R}_{EO}$ and $\mathcal{R}_{CO_2}$, associated with the key species EO and CO_2. These are written in the conventional form in Eqs. (F1a, b) in Fig. 1.2. The relationship (1.9) for the components E, W and O_2 arises from the coefficients of $\mathcal{R}_{EO}$ and $\mathcal{R}_{CO_2}$ (Eq. F2 in Fig. 1.2). Likewise, the σ_{kj} values can be identified directly in the upper right submatrix of ν^r, coloured for this purpose in Fig. 1.2.

The sequence of operations in the manner described to determine the canonical reactions $\mathcal{R}_{EO}$ and $\mathcal{R}_{CO_2}$ only involved the original reactions $\mathcal{R}_1$ and $\mathcal{R}_2$. Then, it can be inferred that $\mathcal{R}_{EO}$ and $\mathcal{R}_{CO_2}$ are linear combinations of only $\mathcal{R}_1$ and $\mathcal{R}_2$. Indeed, it can be concluded from Figs. 1.1 and 1.2 that $\mathcal{R}_{EO} = \mathcal{R}_1$, and $\mathcal{R}_{CO_2} = \frac{1}{2}\mathcal{R}_1 + \frac{1}{2}\mathcal{R}_2$.

It is recalled that the rates r_{EO} and r_{CO_2} of the key species must be evaluated from the original system (1.24), $r_{EO} = r_1 - r_2$, $r_{CO_2} = 2r_2 + 2r_3$. Although only $K = 2$ of the reactions were identified as LI, the rates of **all** the original reactions (r_1, r_2 and r_3) will determine the composition changes of the system. It should also be noted that, if any of the canonical reactions coincide with any of the reactions in the original set, their rates will differ, in general. For example, $\mathcal{R}_{EO} = \mathcal{R}_1$, but $r_{EO} = r_1 - r_2 \neq r_1$.

$$
\begin{array}{c} \\ \mathbf{1} \\ \nu\text{:}\ \mathbf{2} \\ \mathbf{3} \end{array}
\begin{array}{c} \begin{array}{ccccc} CO_2 & W & E & EO & O_2 \end{array} \\ \begin{bmatrix} 0 & 0 & -1 & 1 & -\frac{1}{2} \\ 2 & 2 & 0 & -1 & -\frac{5}{2} \\ 2 & 2 & -1 & 0 & -3 \end{bmatrix} \end{array}
\xrightarrow{\text{Exchanging rows 1 and 2}}
\begin{array}{c} \\ \mathbf{2} \\ \mathbf{1} \\ \mathbf{3} \end{array}
\begin{array}{c} \begin{array}{ccccc} CO_2 & W & E & EO & O_2 \end{array} \\ \begin{bmatrix} 2 & 2 & 0 & -1 & -\frac{5}{2} \\ 0 & 0 & -1 & 1 & -\frac{1}{2} \\ 2 & 2 & -1 & 0 & -3 \end{bmatrix} \end{array}
\xrightarrow{\text{Operating on column 1}}
$$

$$
\begin{array}{c} \\ \mathbf{2} \\ \nu^I\text{:}\ \mathbf{1} \\ \mathbf{3} \end{array}
\begin{array}{c} \begin{array}{ccccc} CO_2 & W & E & EO & O_2 \end{array} \\ \begin{bmatrix} 1 & 1 & 0 & -\frac{1}{2} & -\frac{5}{4} \\ 0 & 0 & -1 & 1 & -\frac{1}{2} \\ 0 & 0 & -1 & 1 & -\frac{1}{2} \end{bmatrix} \end{array}
\xrightarrow{\text{After GJ reduction}}
\begin{array}{c} \\ \mathbf{2} \\ \nu^r\text{:}\ \mathbf{1} \\ \mathbf{3} \end{array}
\begin{array}{c} \begin{array}{ccccc} CO_2 & E & W & EO & O_2 \end{array} \\ \begin{bmatrix} 1 & 0 & 1 & -\frac{1}{2} & -\frac{5}{4} \\ 0 & 1 & 0 & -1 & \frac{1}{2} \\ 0 & 0 & 0 & 0 & 0 \end{bmatrix} \end{array}
$$

Canonical reactions

$$\mathcal{R}_{CO_2} = \mathbf{CO_2} + 1W - \tfrac{1}{2}EO - \tfrac{5}{4}O_2 = 0, \quad \text{or} \quad \tfrac{1}{2}EO + \tfrac{5}{4}O_2 \Leftrightarrow \mathbf{CO_2} + W \qquad \text{(F3.a)}$$

$$\mathcal{R}_{E} = \mathbf{E} - 0W - 1EO + \tfrac{1}{2}O_2 = 0, \quad \text{or} \quad EO \Leftrightarrow \mathbf{E} + \tfrac{1}{2}O_2 \qquad \text{(F3.b)}$$

Relationships between net generation rates

$$r_W = 1r_{CO_2} + 0r_E; \quad r_{EO} = -\tfrac{1}{2}r_{CO_2} - 1r_E; \quad r_{O_2} = -\tfrac{5}{4}r_{CO_2} + \tfrac{1}{2}r_E; \qquad \text{(F4)}$$

Fig. 1.3 Example of GJ reduction with the first columns of ν corresponding to CO_2 and W

Although the ordering of rows and columns of the matrix ν does not change its rank, it is clear that the GJ reduction carried out sequentially from the first column will lead to identifying the species in the first columns as key species. For the example in Fig. 1.1, all pairs except CO_2 and W can fulfil the role of key species. To visualise this fact, ν is arranged with its first two columns corresponding to CO_2 and W, as indicated in the configuration of ν in Fig. 1.3.

To start the reduction of ν by operating from the first column, it is necessary to shift the position of either the first row or first column,[1] as $\nu_{1,1} = 0$. The shift of the latter should be avoided to preserve the intention that CO_2 is a key species. Then, the first two rows/reactions are interchanged. Such an exchange, like any other involving rows or columns, should be recorded in the labels, as indicated in the present case in Fig. 1.3. When the reduction of the first column is completed, the matrix ν^I in Fig. 1.3 results. As $\nu^c_{2,2} = 0$, a new exchange must be performed to continue. In order to maintain the progress achieved in the reduction, the exchange must be made with a **row below** or a **column to the right** of the null coefficient. But in the example, the only lower element is $\nu^c_{3,2} = 0$, so the desirable exchange between the second and third rows does not solve the problem. To proceed, it is only possible to exchange the second column with one to the right of it, which excludes the possibility that W behaves as a key species. By exchanging,

[1] The GJ algorithm as typically used in solving linear systems of equations do not perform column shift as a normal operation, as this implies a reordering of the variables. However, for the purpose of finding the linear combination coefficients of the relationships given in Eq. (1.9), as in the proposed procedure, the column shift facilitates the interpretation of the results.

for example, second and third columns, the procedure can be continued. After proceeding with the resulting second column, the final reduced matrix ν^r is already reached (Fig. 1.3).

As it should be, the last row is null, corresponding to the already known combination $\mathcal{R}_3 = \mathcal{R}_1 + \mathcal{R}_2$. The canonical reactions of the resulting key species CO_2 and E and the relationships (1.9) for the component species W, EO, and O_2 are given in Eqs. (F3a, b) and (F4) of Fig. 1.3. The canonical reactions are related to the original set (by inspection) in the form $\mathcal{R}_{CO_2} = \frac{1}{2}\mathcal{R}_2$, $\mathcal{R}_E = -\mathcal{R}_1$.

The example in Fig. 1.3 covers all practical alternatives that may arise during GJ reduction and allows visualising, that, in general, not all sets of K species will be able to be grouped as key species. It can be mentioned that this will occur whenever the columns of the candidate species in the ν matrix are not LI.

The results obtained from the GJ reduction of the ν matrix and the information provided by the coefficients of the reduced ν^r matrix can be formally generalised:

- The rank K of the matrix ν determines the number of key species.
- A particular set of K key species can be chosen, if feasible, by grouping the respective stoichiometric coefficients in the first columns of ν.
- The canonical reactions correspond to the first K rows of ν^r, as follows:

$$\mathcal{R}_k = A_k + \sum_{j=K+1}^{S} \sigma_{kj} A_j = 0; \qquad k = 1, \ldots, K \tag{1.28}$$

- The coefficients σ_{kj} in rows $k = 1, \ldots, K$ and columns $j = K + 1, \ldots, S$ of ν^r determine the relationships (1.9) for the $(S - K)$ component species.
- The K original reactions associated with the rows of the canonical reactions (according to the labels of such rows) form a subset of LI reactions, while the $(R - K)$ original reactions associated with the (eventually) null rows are linearly dependent on the LI subset.

1.4.1 Coefficients of Linear Combination Between Given Reactions

The sequence of operations (1.26a, b) used for the GJ reduction of the ν matrix allows concluding that the set of K canonical reactions arise by linear combination of the set of K original reactions identified as LI, i.e., those of the first rows of the reduced matrix, ν^r, according to the final identification of the labels. For example, for the reduced matrix ν^r in Fig. 1.3:

$$\mathcal{R}_{CO_2} = \Lambda_{CO_2,1}\mathcal{R}_1 + \Lambda_{CO_2,2}\mathcal{R}_2 = 0\mathcal{R}_1 + \frac{1}{2}\mathcal{R}_2 = \frac{1}{2}\mathcal{R}_2 \tag{1.29a}$$

$$\mathcal{R}_{E} = \Lambda_{E,1}\mathcal{R}_1 + \Lambda_{E,2}\mathcal{R}_2 = -1\mathcal{R}_1 + 0\mathcal{R}_2 = -\mathcal{R}_1 \quad (1.29b)$$

The values of the coefficients $\Lambda_{m,i}$ in Eqs. (1.29a, b) have been established by direct comparison between the canonical reactions and the original ones in Fig. 1.1. Furthermore, it can be concluded that each of the last $(R-K)$ identically null rows are the result of the original reaction associated with the specific row being a linear combination of the set of K original reactions identified as LI. For example, in Fig. 1.3:

$$\underline{\mathbf{0}} = \Lambda_{3,1}\mathcal{R}_1 + \Lambda_{3,2}\mathcal{R}_2 + \Lambda_{3,3}\mathcal{R}_3 = -1\mathcal{R}_1 - 1\mathcal{R}_2 + 1\mathcal{R}_3 \quad (1.29c)$$

where $\underline{\mathbf{0}}$ indicates "identically null". The coefficients $\Lambda_{m,i}$ in Eq. (1.29c) were previously identified by inspection, $\mathcal{R}_3 = \mathcal{R}_1 + \mathcal{R}_2$.

The evaluation of the coefficients $\Lambda_{m,i}$ is of practical interest, for example, to evaluate thermodynamic reaction properties (e.g., enthalpy of reaction, Gibbs free-energy of reaction) of linearly dependent reactions in terms of the values for a set of LI reactions. The coefficients $\Lambda_{m,i}$ in Eqs. (1.29a–c) allow expressing such properties of the canonical reactions and of the dependent reaction $\mathcal{R}_3$ in terms of the values of the two original LI reactions $\mathcal{R}_1$ and $\mathcal{R}_2$. However, for more complex systems than that of the example in Fig. 1.1, a systematic procedure is required to evaluate the combination coefficients $\Lambda_{m,i}$. This can be conveniently carried out by recording all combinations of the type (1.26a, b) performed until the GJ reduction is completed. Based on the example in Fig. 1.3, a way to suitably accomplish this task will be described next.

Consider the reaction $\mathcal{R}_1$ (Fig. 1.1) written in the form: $-E - O_2 + EO = \mathcal{R}_1$. Including those species which do not participate in $\mathcal{R}_1$ and expressing $\mathcal{R}_1$ as an elementary combination of the remaining original reactions, it can be written *in extenso*:

$$\mathbf{0}CO_2 + \mathbf{0}W - \mathbf{1}E + \mathbf{1}EO - \frac{1}{2}O_2 = \mathbf{1}\mathcal{R}_1 + \mathbf{0}\mathcal{R}_2 + \mathbf{0}\mathcal{R}_3 \quad (1.30a)$$

Similarly for $\mathcal{R}_2$ and $\mathcal{R}_3$:

$$\mathbf{2}CO_2 + \mathbf{2}W - \mathbf{0}E - \mathbf{1}EO - \frac{5}{2}O_2 = \mathbf{0}\mathcal{R}_1 + \mathbf{1}\mathcal{R}_2 + \mathbf{0}\mathcal{R}_3 \quad (1.30b)$$

$$\mathbf{2}CO_2 + \mathbf{2}W - \mathbf{1}E + \mathbf{0}EO - \mathbf{3}O_2 = \mathbf{0}\mathcal{R}_1 + \mathbf{0}\mathcal{R}_2 + \mathbf{1}\mathcal{R}_3 \quad (1.30c)$$

The species coefficients in Eqs. (1.30a–c) form the ν matrix and the coefficients of the nominal reactions $\mathcal{R}_i$ define an I_R identity matrix ($R \times R$ in general, 3×3 in the example). We denote I_R as *auxiliary matrix*. In Fig. 1.4, $(\nu \vdots I_R)$ is the *extended matrix* for the configuration of ν used in Fig. 1.3, where the "=" sign in the definitions (1.30a–c) has been omitted but is implicitly retained.

The same GJ reduction procedure is carried out on ν but extending the combinations to the rows of the auxiliary matrix. In this way, after each combination between rows,

$$
(\nu \vdots I_R):\quad
\begin{array}{c}\\ \mathbf{1}\\ \mathbf{2}\\ \mathbf{3}\end{array}
\begin{array}{c}
\begin{array}{ccccc} CO_2 & W & E & EO & O_2 \end{array}\\
\begin{bmatrix} 0 & 0 & -1 & 1 & -\frac{1}{2}\\ 2 & 2 & 0 & -1 & -\frac{5}{2}\\ 2 & 2 & -1 & 0 & -3 \end{bmatrix}
\end{array}
\begin{array}{c}
\begin{array}{ccc} \mathcal{R}_1 & \mathcal{R}_2 & \mathcal{R}_3 \end{array}\\
\begin{bmatrix} 1 & 0 & 0\\ 0 & 1 & 0\\ 0 & 0 & 1 \end{bmatrix}
\end{array}
\xrightarrow{\text{After GJ reduction}}
$$

$$
(\nu^r \vdots M_{GJ}):\quad
\begin{array}{c}\\ \mathbf{2}\\ \mathbf{1}\\ \mathbf{3}\end{array}
\begin{array}{c}
\begin{array}{ccccc} CO_2 & E & W & EO & O_2 \end{array}\\
\begin{bmatrix} 1 & 0 & 1 & -\frac{1}{2} & -\frac{5}{4}\\ 0 & 1 & 0 & -1 & \frac{1}{2}\\ 0 & 0 & 0 & 0 & 0 \end{bmatrix}
\end{array}
\begin{array}{c}
\begin{array}{ccc} \mathcal{R}_1 & \mathcal{R}_2 & \mathcal{R}_3 \end{array}\\
\begin{bmatrix} 0 & \frac{1}{2} & 0\\ -1 & 0 & 0\\ -1 & -1 & 1 \end{bmatrix}
\end{array}
$$

Fig. 1.4 Example of GJ reduction of an extended matrix $(\nu \vdots I_R)$

the coefficients of the auxiliary matrix rows are also updated, indicating how the new reactions are related to the original reactions $\mathcal{R}_i$. After the GJ reduction is completed, the transformed extended matrix $(\nu^r \vdots M_{GJ})$ is obtained, as shown in Fig. 1.4, where M_{GJ} is the transformed matrix from I_R. It is convenient to remember that all operations on the rows must be extended to the auxiliary matrix; thus, if the procedure requires exchanging rows (e.g., between rows 1 and 2 in the example), the rows of the auxiliary matrix must also be exchanged. Each final reaction in the rows of ν^r corresponds to a linear combination of the original reactions $\mathcal{R}_i$, defined by combination coefficients $\Lambda_{m,i}$ matching the coefficients of the matrix M_{GJ} in the same row. For the example, it is found that Eqs. (1.29a–c) results.

We can generalise the procedure and conclusions to **any** reaction system. After the GJ reduction of the extended matrix $(\nu \vdots I_R)$, the reduced matrix $(\nu^r \vdots M_{GJ})$ allows us to identify the following combinations:

$$\text{In the first } K \text{ rows (canonical reactions)}: \quad \mathcal{R}_k = \sum_{i=1}^{K} \Lambda_{k,i}\mathcal{R}_i \tag{1.31a}$$

$$\text{In the last } (R-K) \text{ rows}: \quad \underline{\mathbf{0}} = \sum_{i=1}^{K} \Lambda_{m,i}\mathcal{R}_i + \Lambda_{m,m}\mathcal{R}_m \tag{1.31b}$$

where $\mathcal{R}_k$ are canonical reactions, $\mathcal{R}_i$ are the original K reactions identified as LI, and $\mathcal{R}_m$ are the $(R-K)$ original reactions identified as linearly dependent on the set $\{\mathcal{R}_i\}$ of LI reactions. Coefficients $\Lambda_{m,i}$ for each of Eqs. (1.31a, b), are the elements of the corresponding rows of the reduced auxiliary matrix M_{GJ}.

By definition of the combinations (1.26a, b), $\Lambda_{m,m} \neq 0$ will always hold in Eq. (1.31b). Then, it can be written:

$$\mathcal{R}_m = \sum_{i=1}^{K} \lambda_{m,i}\mathcal{R}_i; \quad \lambda_{m,i} = -\Lambda_{m,i}/\Lambda_{m,m} \tag{1.31c}$$

The above procedure can be used if the sole purpose is to identify a subset of K linearly independent reactions from R given reactions, the complementary subset of $(R - K)$ linearly dependent reactions and the values of linear combination coefficients $\Lambda_{m,i}$. Those reactions which are known or presumed to be LI must be arranged in the upper rows of ν. If it happens that an element on the main diagonal is zero, in order to proceed with the GJ reduction, the exchange of the corresponding row with a lower one should be avoided, as far as possible, by performing instead an exchange between columns of the species. When the reduction is completed, the procedure will confirm whether a presumed subset is indeed LI or not.

Finally, it is noted that, if the definition of a set of key species is not required, a Gaussian reduction (reduction to the form of an *upper triangular matrix*) will suffice to identify the dependent reactions and the corresponding linear combination coefficients (Eq. 1.31b).

1.4.2 Application to Reaction Mechanisms

Consider a given mechanism of R_{ERS} elementary reaction steps involving *reaction intermediates* (RI), defined as species with negligible net production rates after an initial period much shorter than the time scale of the *observable species* changes (pseudo-steady state hypothesis). Based on this concept, the net production rates r_j of the observable species will not depend on the concentrations of the RI. If the mechanism corresponds to a single reaction, the rates r_j of the observable species will satisfy Eqs. (1.2a, b) and thus the *global reaction* stoichiometry can be directly identified. Instead, if the mechanism involves multiple global reactions, a procedure must be implemented to specify a representative set of global LI reactions.

To this end, let the matrix ν for the given mechanism be defined by rows and columns corresponding to all elementary steps and all RI and observable species, respectively. As the global reactions should not include any RI, the stoichiometric coefficients of the RI should be arranged in the first columns of ν, so that the GJ reduction of ν can lead to identify all RI as key species along with their respective canonical reactions. In this way, the RI will not participate in the rest of the canonical reactions, which will therefore correspond to a set of global LI reactions.

As a particular situation that may occur, some of the RI may not be able to remain as key species after the GJ reduction. In such a case, they should **only** participate in the canonical reactions of the rest of the RI. If this does not occur, the conclusion will be that the proposed mechanism is not consistent with the assumed identification of the RI. On the other hand, if an observable species behaves as an inhibitor or a catalyst of the observable reactions, it cannot be a key species in any of the global canonical reactions and will only participate in the canonical reactions of the RI.

The reduced matrix ν^r will have K_{RI} rows/reactions with RI as key species, K rows/reactions with observable species as key species and, eventually, $(R_{ERS} - K_{RI} - K)$ identically null rows, which means that the corresponding elementary steps are linear combinations of the remaining $(K_{RI}+K)$ steps. This last situation will respond to the fact that the identified global reactions can be represented by different sequences of elementary steps, called *reaction pathways*. Their significance, particularly for the development of kinetic expressions from solid-catalysed reaction mechanisms, can be found in Constalcs et al. [29] and Marin et al. [30], and will not be addressed here.

As an example, we consider the hypothetical four-stage mechanism defined in Fig. 1.5, with five observable species (A, B, C, D, Q) and two RI (A^* and A^*B). The chosen configuration for ν is also presented in Fig. 1.5, where the coefficients of C have been placed in the third column, with the intention that C may result a key species. During the reduction, however, it is necessary to exchange the C column twice. The reduced matrix ν^r is shown in Fig. 1.5.

The resulting key species are A^*, A^*B, D and Q, while C could not remain as such. There are $K = 2$ global canonical reactions with D and Q as key species (in the third and fourth rows of ν^r). The relationships of type (1.9) for the component species C, A, and B, and the results after making $r_{A^*} = r_{A^*B} = 0$, are also given in Fig. 1.5. Since $r_C = 0$, it is concluded that C plays the role of catalyst or inhibitor, an effect to be determined once the overall reaction rates are expressed from the rates of the elementary stages and the conditions $r_{A^*} = r_{A^*B} = 0$. In the present example, C turns out to behave as a catalyst.

Step 1: $A + C \Leftrightarrow A^*$
Step 2: $A^* + B \Leftrightarrow A^*B$
Step 3: $A^*B + A \Leftrightarrow C + D$
Step 4: $A^*B + B \Leftrightarrow C + Q$

$$\Rightarrow \nu: \begin{array}{c} \\ 1 \\ 2 \\ 3 \\ 4 \end{array} \begin{array}{ccccccc} A^* & A^*B & C & D & Q & A & B \\ \end{array}$$
$$\nu = \begin{bmatrix} 1 & 0 & -1 & 0 & 0 & -1 & 0 \\ -1 & 1 & 0 & 0 & 0 & 0 & -1 \\ 0 & -1 & 1 & 1 & 0 & -1 & 0 \\ 0 & -1 & 1 & 0 & 1 & 0 & -1 \end{bmatrix} \xrightarrow{\text{After GJ reduction}}$$

(columns of ν: A^*, A^*B, C, D, Q, A, B; rows 1–4)

$$\nu^r = \begin{bmatrix} 1 & 0 & 0 & 0 & -1 & -1 & 0 \\ 0 & 1 & 0 & 0 & -1 & -1 & -1 \\ 0 & 0 & 1 & 0 & 0 & -2 & -1 \\ 0 & 0 & 0 & 1 & 0 & -1 & -2 \end{bmatrix}$$

(columns of ν^r: A^*, A^*B, D, Q, C, A, B; rows 1–4)

Observable canonical reactions:

$\mathcal{R}_D = \mathbf{D} - 2A - 1B = 0$, or: $2A + B \Leftrightarrow \mathbf{D}$

$\mathcal{R}_Q = \mathbf{Q} - 1A - 2B = 0$, or: $A + 2B \Leftrightarrow \mathbf{Q}$

Relationships between net generation rates

$$r_C = -1r_{A^*} - 1r_{A^*B};$$
$$r_A = -1r_{A^*} - 1r_{A^*B} - 2r_D - 1r_Q;$$
$$r_B = 0r_{A^*} - 1r_{A^*B} - 1r_D - 2r_Q;$$

After making $r_{A^*} = r_{A^*B} = 0$:

$r_C = 0$ (**C is a catalyst/inhibitor**); $r_A = -2r_D - 1r_Q$; $r_B = -1r_D - 2r_Q$;

Fig. 1.5 Example of the GJ reduction of a matrix ν, corresponding to a reaction mechanism

1.4.3 Possible Reactions in a Mixture of Reacting Species

Section 1.3.3 described the problem of specifying a set of reactions from the experimental identification of certain reacting species. An extension of the procedure in Sect. 1.4.2 can be used as an alternative to determine the maximum possible number of LI reactions and a feasible set of canonical reactions.

Consider as an example a system in which variations in the amounts of $S = 4$ species are observed: ethylene oxide (EO: C_2H_4O), ethylene glycol (EG: $C_2H_6O_2$), diethylene glycol (DEG: $C_4H_{10}O_3$) and water (W: H_2O). The *formation reactions* of these species from the elements in their atomic form are:

$$\mathcal{R}_{f,EO} = EO - 2C - 4H - O = 0$$

$$\mathcal{R}_{f,EG} = EG - 2C - 6H - 2O = 0$$

$$\mathcal{R}_{f,DEG} = DEG - 4C - 10H - 3O = 0$$

$$\mathcal{R}_{f,W} = W - 2H - O = 0$$

The idea is to assimilate the formation reactions to elementary steps of a hypothetical reaction mechanism, assigning to the elements (C, H and O in the example) the role of "reaction intermediates", thus considering that they are not "observed" in the reacting system.

Before continuing, it is useful to clarify two aspects not involved in the considered example. The first one is the presence of a molecular species consisting of a single element. In this case, its formation reaction from the atomic element should also be included (for example, if H_2 is a reacting species, $\mathcal{R}_{f,H_2} = H_2 - 2H = 0$). The second aspect is when there are electrically charged species. In this situation, the total number E of elements, or "reaction intermediates", must include as an additional "element" the electric charge, with positive values for protons and negative values for anions. For example, denoting one mole of elementary electric charge as Q, if $SO_3^=$ is a reacting species, $\mathcal{R}_{f,SO_3^=} = SO_3^= - (S + 3O - 2Q) = 0$.

The stoichiometric matrix ν for the formation reactions is now built up. The first columns are assigned to the "reaction intermediates" and the next following columns to the observable species intended to behave as key species. For the example, the following configuration is adopted:

$$\nu: \begin{array}{c} \\ \mathbf{W} \\ \mathbf{DEG} \\ \mathbf{EO} \\ \mathbf{EG} \end{array} \begin{array}{c} \begin{array}{ccccccccc} C & H & O & & W & DEG & EO & EG \end{array} \\ \left[\begin{array}{ccccccc} 0 & -2 & -1 & \vdots & 1 & 0 & 0 & 0 \\ -4 & -10 & -3 & \vdots & 0 & 1 & 0 & 0 \\ -2 & -4 & -1 & \vdots & 0 & 0 & 1 & 0 \\ -2 & -6 & -2 & \vdots & 0 & 0 & 0 & 1 \end{array} \right] \end{array} \quad (1.32)$$

When proceeding with the GJ reduction of the matrix ν, the "reaction intermediate" O cannot be retained as a key species, so the reduced matrix results:

$$\nu^r: \quad \begin{array}{c} \\ \mathbf{EO} \\ \mathbf{EG} \\ \\ \mathbf{W} \\ \mathbf{DEG} \end{array} \begin{array}{c} \begin{array}{ccccccc} C & H & W & DEG & O & EO & EG \end{array} \\ \left[\begin{array}{cc|ccccc} 1 & 0 & 0 & 0 & -\frac{1}{2} & -\frac{3}{2} & 1 \\ 0 & 1 & 0 & 0 & \frac{1}{2} & \frac{1}{2} & -\frac{1}{2} \\ \hline 0 & 0 & 1 & 0 & 0 & 1 & -1 \\ 0 & 0 & 0 & 1 & 0 & -1 & -1 \end{array}\right] \end{array} \tag{1.33}$$

It can be noted that the first two rows correspond to the canonical reactions for the generation of the elements C and H from the component species EO, EG and the element O. The last two rows/reactions involved **only** the observable species and correspond to the **canonical reactions** for W and DEG:

$$\mathcal{R}_W = \mathbf{W} + EO - EG = 0; \qquad \mathcal{R}_{DEG} = \mathbf{DEG} - EO - EG = 0$$

Written as in the actual production of ethylene glycol:

$$\mathbf{W} + EO \leftrightarrow EG; \qquad EO + EG \leftrightarrow \mathbf{DEG}$$

In Sect. 1.3.3, the atomic matrix β was defined as a matrix with each row associated with an element and coefficients corresponding to the number of atoms of that element in each of the reacting species. In the example, $(-\beta^T)$ corresponds to the left submatrix in (1.32):

$$\beta^T: \quad \begin{array}{c} \\ \mathbf{W} \\ \mathbf{DEG} \\ \mathbf{EO} \\ \mathbf{EG} \end{array} \begin{array}{c} \begin{array}{ccc} C & H & O \end{array} \\ \left[\begin{array}{ccc} 0 & 2 & 1 \\ 4 & 10 & 3 \\ 2 & 4 & 1 \\ 2 & 6 & 2 \end{array}\right] \end{array}$$

In general, the rank of β^T (equal to that of β) will be $N_\beta \leq E$. In the example, $N_\beta = 2 < E = 3$, so the element O could not play the role of a key species. Whenever it happens that $N_\beta < E$, the elements that cannot be kept as keys species will only participate in the canonical reactions of the remaining elements, as follows from Eq. (1.33) in the example. In addition, as by construction the rank of ν always equals S, the number of global LI reactions will be $K = S - N_\beta$. In the example, $K = 2$.

Finally, it should be noted that, given the reacting species, the procedure outlined in Sect. 1.3.3 to define a feasible set of LI reactions is computationally more efficient. Nonetheless, the alternative presented here maintains the same strategy followed throughout Sect. 1.4 to achieve different purposes: resorting to linear combinations of reactions or elementary reaction steps.

1.4.4 Comments on the Adoption of the Proposed Procedure in a CRE Course

Concerning the primary goal of evaluating the coefficients σ_{kj} of the relationships (1.9), a classroom presentation can cover the concepts in Sect. 1.4 up to the point of recognising that the search for a set of canonical reactions is assimilated to the GJ reduction of the matrix ν. Depending on the perception about the students' background, details on the GJ reduction procedure may be omitted, albeit the results in Figs. 1.2 and 1.3 should be invoked to summarise the general conclusions. Regarding the practice of the GJ reduction, Appendix 2 presents a series of assignments for an interactive session under MATLAB or Octave platforms, which are mainly intended to avoid manual updating of the intermediate matrices, while at the same time allow recalling the step sequence and decisions (as for exchanging rows or columns) until obtaining the reduced matrix.

Beyond the initial objective of evaluating the coefficients σ_{kj} in Eq. (1.9), the material presented in Sect. 1.4.1–1.4.3 illustrates additional applications of the same procedure based on the concept of linear combination of reactions.

The evaluation of the coefficients of linear combination between given reactions discussed in Sect. 1.4.1 deserves special consideration. Although the concept of canonical reactions was used for the evaluation of coefficients σ_{kj}, such a concept is not indispensable to study the behaviour of isothermal reactors with ideal flow patterns. However, when considering adiabatic operations, the enthalpy of the mixture remains constant, which implies a relationship between temperature and concentration variations. As the latter can be expressed only in terms of the variation of the key species, a relationship between temperature change and key species concentrations variations ultimately arises. Such a relationship will be considered in Chaps. 2 and 3 in a more general way, but in this instance, it is sufficient to consider the case where the density (ρ), heats of reaction (ΔH_k), and specific heat ($\hat{c}_p$) remain constant, a situation in which Eq. (1.34) results:

$$T - T_0 = \sum_{k=1}^{K} [(-\Delta H_k)/(\rho \hat{c}_p)](C_k - C_{k0}), \tag{1.34}$$

where $\Delta H_k = h_k + \sum_{j=K+1}^{S} \sigma_{kj} h_j$ is the heat of reaction of the canonical reaction $\mathcal{R}_k$ and h_j is the partial molar enthalpy of species A_j. Then, the use of (1.34) requires invoking the concept of canonical reactions. For this purpose, considering the general case that the original reaction set $\{\mathcal{R}_i\}$ is not in the form of canonical reactions, it is reasonable to assume that values of ΔH_i are given, instead of ΔH_k. In such a case, the coefficients $\Lambda_{k,i}$ evaluated in Sect. 1.4.1 will be required for the evaluation of the latter in the form:

$$\Delta H_k = \sum_{k=1}^{K} \Lambda_{k,i} \Delta H_i \tag{1.35}$$

where a subset of K linearly independent reactions from the original set $\{\mathcal{R}_i\}$ is involved.

1.5 Conclusions

From the beginning of this chapter, the existence of stoichiometric relationships between the net production rates, r_j, of the species (Eq. 1.9) for multiple reactions systems has been emphasised. Such relationships are a generalisation of those holding for single reactions (see Eq. 1.2a, b) and as such allow the analysis and study of all types of reacting systems with the necessary stoichiometric restrictions.

Although the relationships between the rates r_j for a single reaction are invoked and recognised as a basic property in the study of reaction systems in introductory CRE courses or in the corresponding textbooks, their extension to the case of multiple reactions is not frequent. This is most likely due to the prior elaboration required to establish such relationships (Eq. 1.9) for a given set of reactions. The procedures to this end in the literature have been discussed in Sect. 1.3, where it becomes evident that they require close familiarity with various concepts and operations of lineal algebra. While students of an introductory CRE course can be expected to have knowledge on that matter, the application, contextualisation and exercitation to the problem under consideration will require a significant students' effort and time allotted in the course schedule.

In Sect. 1.4, a procedure that attempts to minimise the use of linear algebra to establish the relationships between the rates r_j has been proposed. It is based on two facts. First, the recognition that the coefficients σ_{kj} in Eq. (1.9) can be directly evaluated for a set of canonical reactions. The second fact is that a given set of multiple reactions can be represented in different ways from proper linear combinations that keep unchanged the stoichiometric relationships between the net production rates of the species. Then, by means of a suitable example, it is shown that an ordered sequence of linear combinations allows transforming the original reaction set into a set of canonical reactions. The example makes it possible to recognise that such a succession of combinations is equivalent to the Gauss–Jordan (GJ) reduction of the original stoichiometric matrix ν. This leads to generalise the procedure and conclusions, based on invoking the relationship between the GJ reduction and the concept of the rank of a matrix. In this way, the suggested procedure only resorts to elementary matrix properties.

The linear combination of reactions used as the basis of the procedure provides as a corollary the introduction of the concept of linear independence of reactions and, by the use of an extended stoichiometric matrix, the determination of the combination coefficients (Sect. 1.4.1). It is finally shown that the concept of linear combination of reactions can also be conveniently employed for the analysis of reaction mechanisms (Sect. 1.4.2) and the specification of possible reactions from the identification of reacting species (Sect. 1.4.3).

Appendix 1: On the Independence of the Production Rates of the Species Under the Linear Combination of Reactions

Consider as a control case a set of three reactions, $\mathcal{R}_1$, $\mathcal{R}_2$ and $\mathcal{R}_3$, for which the combination (1.26a, b) is applied to replace $\mathcal{R}_3$ by a new reaction $\mathcal{R}_3^n$ from the linear combination with $\mathcal{R}_2$, i.e., $m = 3$ and $l = 2$. According to Eq. (1.26b), the coefficients of $\mathcal{R}_3^n$ for arbitrary values α_2 and $\alpha_3 \neq 0$ result:

$$\nu_{3j}^n = \alpha_3 \nu_{3j} + \alpha_2 \nu_{2j}; \qquad j = 1, \ldots, S \tag{1.36}$$

For the original set $\{\mathcal{R}_1, \mathcal{R}_2, \mathcal{R}_3\}$, the generation rates $\mathscr{r}_j$ are expressed as:

$$\mathscr{r}_j = \nu_{1j} r_1 + \nu_{2j} r_2 + \nu_{3j} r_3; \qquad j = 1, \ldots, S \tag{1.37a}$$

The reaction rates of the new set $\{\mathcal{R}_1, \mathcal{R}_2, \mathcal{R}_3^n\}$ must be reassigned with new values r_i^n $(i = 1, 2, 3)$. Then, for each $\mathscr{r}_j$,

$$\mathscr{r}_j = \nu_{1j} r_1^n + \nu_{2j} r_2^n + \nu_{3j}^n r_3^n; \qquad j = 1, \ldots, S \tag{1.37b}$$

The values r_i^n must be defined in such a way as to maintain the same values $\mathscr{r}_j$. Then, subtracting Eqs. (1.37a, b), it must be verified that:

$$0 = \nu_{1j}(r_1 - r_1^n) + \nu_{2j}(r_2 - r_2^n) + \nu_{3j} r_3 - \nu_{3j}^n r_3^n; \qquad j = 1, \ldots, S$$

Replacing ν_{3j}^n from Eq. (1.36) and grouping terms in ν_{1j}, ν_{2j}, ν_{3j}:

$$0 = \nu_{1j}(r_1 - r_1^n) + \nu_{2j}(r_2 - r_2^n - \alpha_2 r_3^n) + \nu_{3j}(r_3 - \alpha_3 r_3^n); \quad j = 1, \ldots, S \tag{1.38}$$

Equation (1.38) must hold for any set of values r_i $(i = 1, 2, 3)$, which requires that the three factors multiplying ν_{1j}, ν_{2j} and ν_{3j} must be set to zero. Thus, $r_1^n = r_1$, $r_2^n = r_2 - \alpha_2 r_3^n$, $r_3^n = r_3/\alpha_3$. Replacing the value r_3^n in the expression for r_2^n, we get the new r_i^n values to be assigned:

$$r_1^n = r_1 \tag{1.39a}$$

$$r_2^n = r_2 - (\alpha_2/\alpha_3) r_3 \tag{1.39b}$$

$$r_3^n = r_3/\alpha_3 \tag{1.39c}$$

The values r_i^n needed to maintain the rates r_j are specified only in terms of the values of the coefficients α_2 and α_3. It becomes clear that $\alpha_3 \neq 0$ must effectively hold. It is straightforward to extend the conclusions to any number R of original reactions. Those reactions that do not participate in the combination (1.26a, b) will retain the same values of their reaction rates (as happen in the example with r_1 for $\mathcal{R}_1$), while for the substituted reaction and for the auxiliary reaction ($\mathcal{R}_2$ and $\mathcal{R}_3$, respectively, in the given example), the equivalence with Eqs. (1.39b, c) is maintained. Explicitly:

$$r_m^n = r_m^c / \alpha_m, \tag{1.40a}$$

$$r_l^n = r_l^c - \left(\frac{\alpha_l}{\alpha_m}\right) r_m^c, \tag{1.40b}$$

$$r_i^n = r_i^c; \qquad i \neq m, l \tag{1.40c}$$

Appendix 2: A Set of Assignments for the GJ Reduction of the Matrix ν Under MATLAB or Octave Platforms

Computational software such as MATLAB, Wolfram Mathematica, Octave, etc. have specific routines for performing the GJ reduction. However, for teaching purposes, it is desirable to have a step-by-step procedure to carry out the operation, allowing intermediate results to be appreciated and decisions to be made when exchange between columns or rows is required. To that end, a set of assignments (in *italics* in the list below) is described here for execution in interactive mode in either MATLAB or Octave.

After the execution of each of the assignments to be described, the resulting configuration for the transformation of ν is automatically displayed. Hence, it allows deciding which assignment is to be executed next.

The stoichiometric matrix ν with R rows and S columns is first defined. Optionally, ν can be defined as the extended matrix $(\nu \vdots I_R)$, to determine the linear combination coefficients between reactions (Sect. 1.4.1).

The index p for column reduction is initialised and the matrix ν (written as *v* below) is extended in the last command (optional):

p = 1;   [R,S] = size(v);   v = [v, eye(R, R)]

- If the current value of $\nu(p, p)$ verifies $\nu(p, p) \neq 0$, the transformation assignments are executed: row p is transformed to achieve $\nu(p, p) = 1$, column p is reduced and finally the index p is updated:

v(p,:) = v(p,:)/v(p,p)

*for k = 1:R; if k =p; v(k,:) = v(k,:) - v(p,:)*v(k,p); end; end; v, p = p + 1*

If the updated index results in $p = R + 1$ or if all rows with index p or higher are identically zero, the procedure stops, with the rank of the matrix ν determined by $K = p - 1$.

- If $\nu(p, p) = 0$, column or row exchange is attempted, depending on the intention to keep the sequence of either the former or the latter.

A column exchange will be feasible if $\nu(p, t) \neq 0$ for at least some value t such that $p < t \leq S$. If possible, a value t is chosen, and the exchange is executed:

v(:,[p,t]) = v(:,[t,p])

The values of p and t for the exchanged columns are manually recorded and the transformation assignments are executed.

A row exchange will be feasible if $\nu(q, p) \neq 0$ for at least some value q such that $p < q \leq R$. If possible, a value q is chosen, and the exchange is executed:

v([p,q],:) = v([q,p],:)

The values of q and t for the exchanged files are recorded and the transformation assignments are executed.

From the final matrix ν, displayed on the screen, the first K rows and columns correspond to the identity matrix of $K \times K$, the following $S-K$ columns contain the coefficients σ_{kj} and, if the extended matrix $\nu \equiv [\nu \vdots I_R]$ was used, the remaining R columns contain the coefficients $\Lambda_{m,i}$ of linear combinations. The record of exchanged rows and/or columns allows the identification of the species in the final columns and the reactions in the final rows.

The same assignments and decisions may be applied for the case of reaction mechanisms (Sect. 1.4.2) and for the determination of feasible reactions when reacting species have been identified (Sect. 1.4.3).

References

1. Walas, S. M. (1959). *Reaction kinetics for chemical engineers*. McGraw-Hill.
2. Carberry, J. J. (1976). *Chemical and catalytic reaction engineering*. McGraw-Hill.

3. Smith, J. M. (1981). *Chemical engineering kinetics*. McGraw-Hill.
4. Westerterp, K. R., Van Swaaij, W. P. M., & Beenackers, A. A. C. M. (1984). *Chemical reactor design and operation* (2nd ed.). Wiley.
5. Holland, C. D., & Anthony, R. G. (1989). *Fundamentals of chemical reaction engineering* (2nd ed.). Prentice-Hall.
6. Schmidt, L. D. (1998). *The engineering of chemical reactions*. Oxford University Press.
7. Levenspiel, O. (1996). *The chemical reactor omnibook*. OSU Book Stores Inc.
8. Levenspiel, O. (1998). *Chemical reaction engineering* (3rd ed.). Wiley.
9. Santamaría, J. M., Herguido, J., Menéndez, M. Á., & Monzón, A. (2002). *Ingeniería de Reactores*. Síntesis.
10. Fogler, H. S. (2016). *Elements of chemical reaction engineering* (5th ed.). Prentice-Hall.
11. Aris, R. (1965). *Introduction to the analysis of chemical reactors*. Prentice-Hall.
12. Aris, R. (1969). *Elementary chemical reactor analysis*. Prentice-Hall.
13. Villermaux, J. (1993). *Génie de la réaction chimique. Conception et fonctionnement des réacteurs* (2nd ed.). TEC&DOC.
14. Hill, C. G. J. (1977). *An introduction to chemical engineering kinetics & reactor design*. Wiley.
15. Trambouze, P., Van Landeghem, H., & Wauquier, J. P. (1984). *Les réacteurs chimiques*. Éditions Technip.
16. Farina, I. H., Ferreti, O. A., & Barreto, G. F. (1986). *Introducción al Diseño de Reactores Químicos*. EUDEBA.
17. Nauman, E. B. (2002). *Handbook of chemical reactor design*. McGraw-Hill.
18. Missen, R. W., Mims, C. A., & Saville, B. A. (1999). *Introduction to chemical reaction engineering and kinetics*. Wiley.
19. Salmi, T. O., Mikkola, J.-P., & Wärnå, J. P. (2011). *Chemical reaction engineering and reactor technology*. CRC Press.
20. Doraiswamy, L. K., & Üner, D. (2013). *Chemical reaction engineering: Beyond the fundamentals*. CRC Press.
21. Alves, J. A., Bressa, S. P., Martínez, O. M., & Barreto, G. F. (2012). Kinetic study of the selective catalytic hydrogenation of 1,3-butadiene in a mixture of n-butenes. *Journal of Industrial and Engineering Chemistry, 18*(4), 1353–1365.
22. Petersen, E. E. (1965). *Chemical reaction analysis*. Prentice-Hall.
23. Froment, G. F., Bischoff, K. B., & De Wilde, J. (2011). *Chemical reactor analysis and design* (3rd ed.). Wiley.
24. Björnbom, P. H. (1975). The independent reactions in calculations of complex chemical equilibria. *Industrial and Engineering Chemistry Fundamentals, 14*(2), 102–106.
25. Smith, W. R., & Missen, R. W. (1991). *Chemical reaction equilibrium analysis: theory and algorithms*. Krieger.
26. Smith, W. R., & Missen, R. W. (1979). What is chemical stoichiometry? *Chemical Engineering Education, Winter*(1), 26–32.
27. Björnbom, P. H. (1973). The relation between the reaction mechanism and the stoichiometric behaviour of chemical reactions. *AIChE Journal, 23*(3), 285–288.
28. Rebsdat, S., & Mayer, D. (2001). Ethylene Oxide. In: Ullmann (Ed.), Encyclopedia of industrial chemistry. Wiley-VCH Verlag GmbH & Co. KGaA.
29. Constales, D., Yablonsky, G. S., D'hooge, D. R., Thybaut, J. W., & Marin, G. B. (2017). *Advanced data analysis and modelling in chemical engineering*. Elsevier B.V.
30. Marin, G. B., Yablonsky, G. S., & Constales, D. (2019). *Kinetics of chemical reactions*. Wiley-VCH.

2 Use of Stoichiometric Relationships in Simple Reaction Systems

2.1 Introduction

The use of stoichiometric relationships between the net production rates of species, r_j, for single fluid phase reacting systems is considered throughout this chapter. As commonly found in introductory CRE courses, these systems are modelled with the well-known *ideal-flow* assumptions. The expectation and main goal in this chapter are to obtain relationships between the amounts of the species present in the reacting systems on the basis of the stoichiometric relationships among the net production rates.

Specifically, the following cases are analysed:

(a) Overall mass balances in steady-state flow systems and their application when chemical equilibrium is reached.
(b) Steady-state tubular reactors with plug-flow hypothesis.
(c) Steady-state flow reactors with perfect mixing behaviour.
(d) Closed reactors with perfect mixing (batch reactors).
(e) Open reactors with perfect mixing in transient regime.

Case (a) omits the detail of the evolution of the chemical species inside the reactor. Instead, the net effect of the chemical reactions between the inlet and outlet streams is considered. Therefore, case (a) is relevant in a CRE course for evaluating the overall behaviour of an individual reactor or its impact when integrated into a continuous process plant. The treatment under chemical equilibrium conditions allows including the thermodynamic constraints in such evaluations.

As is emphasised in Sect. 2.2 of the present chapter, stoichiometry relationships can be conveniently stated in terms of the number of moles of the species for cases (a)–(e). In Sects. 2.3–2.8, the possibility of extending such relationships to an appropriate measure

G. F. Barreto and C. D. Luzi, *Applying Multiple-Reaction Stoichiometry to Chemical Reactor Modelling*, Synthesis Lectures on Chemical Engineering and Biochemical Engineering, https://doi.org/10.1007/978-3-031-42375-8_2

of the species concentration is further discussed. The main conclusions are reported in Sect. 2.9.

Finally, although a single fluid phase with homogeneous reactions taking place is considered throughout the chapter, some of the situations to be analysed can be extended to the case of heterogeneous catalytic reactions. Case (a) does not require further consideration. For the tubular reactor in case (b), the formulation of Sect. 2.5 assuming plug-flow behaviour can also be applied for a packed-bed catalytic reactor if the production rates of the species, $\mathscr{r}_j$, are interpreted as effective catalytic rates per unit bed volume. On the other hand, the concept of residence time, as discussed in Sect. 2.5.2 is no longer applicable. Regarding system (c), the treatment for a flow reactor with perfect mixing in steady state will also be valid when there is a suspended solid catalyst in the stirred solution. Cases (d) and (e) correspond to transient operations, when the accumulation of the species inside a porous catalyst will impose a different behaviour that needs special consideration (as discussed in Chap. 3). Thus, the formulation presented in Sects. 2.7 and 2.8 cannot be extended directly.

2.2 Stoichiometric Relationships Between the Quantities of Reacting Species

It is first mentioned that, for compactness of the mathematical expressions in this chapter, the bounds in summation and product operators will be implicitly associated with the index for a certain specific group of species. The most frequently used examples are:

$\sum_i \equiv \sum_{i=1}^{R}$: a sum of some property associated with reactions, over all the R reactions;

$\sum_j \equiv \sum_{j=1}^{S}$: a sum of some property associated with species, over all the S chemical species present in the system;

$\sum_{j>K} \equiv \sum_{j=K+1}^{S}$: a sum of some property associated with species, over all the $(S-K)$ component species;

$\sum_k \equiv \sum_{k=1}^{K}$: a sum of some property associated with species, over all the K key species (index k' may also be used for key species).

For other occasional groupings, their indices and associated extensions will be defined in due course, if necessary.

The possibility of establishing mass balances in terms of extensive variables for each species, such as molar flow rates (F_j) or the number of moles (N_j), is a common characteristic of all situations (a)–(e) listed in the introduction. This is feasible in case (a) because overall balances are used, while the composition is either supposed to vary in a single spatial direction and without dispersion effects (case b) or is assumed to be spatially invariable (cases c–e).

On the one hand, extensive variables will allow the convenient use of stoichiometry relationships to obtain equivalent relationships between them, except for certain limitations arising in case (e). On the other hand, properties of the system depend on intensive variables—particularly reaction rates or chemical equilibrium conditions—. Although those properties can be expressed in terms of molar flow rates (F_j) or number of moles (N_j), it has a drawback from a didactic point of view: it obscures the fact that the system is determined in terms of intensive variables, including time or residence time (or any other equivalent variable) when the reactions are rate limited. Consequently, the convenience of expressing the balances in terms of intensive variables arises.

The use of different alternatives for expressing the composition of the system, with specific emphasis on the possibility of establishing simple stoichiometric relationships between the amounts of the species, will be analysed in the remainder of this section. The pursued objective is to reduce the number of conservation equations and to identify the minimum number of state variables to represent the system evolution.

As a matter of convenience, case (b), *Steady-state tubular reactors with plug-flow hypothesis*, will be considered here. The conclusions will remain valid in general, while specific points will be discussed when dealing with each case (a)–(e).

The conservation equation for each of the S species can be basically expressed in terms of the molar flow rates F_j as:

$$dF_j/dV = r_j \tag{2.1}$$

Equation (2.1) is subjected to, $F_j(0) = F_{j0}$, where the subscript "0" indicates the reactor inlet. It is recalled that the stoichiometric relationships between the net production rates are given by:

$$r_j = \sum_k \sigma_{kj} r_k; \qquad j > K \tag{2.2}$$

Specifically for the key species ($1 \leq k \leq K$), Eq. (2.1) reads

$$dF_k/dV = r_k \tag{2.3}$$

By replacing both r_j for the component species ($j > K$) from Eq. (2.1) and r_k for the key species from Eq. (2.3) in (2.2), it is obtained $dF_j/dV = \sum_k \sigma_{kj} dF_k/dV$ ($j > K$). Upon integration from the reactor inlet up to a generic position:

$$F_j = F_{j0} + \sum_k \sigma_{kj}(F_k - F_{k0}); \qquad j > K \tag{2.4}$$

Local values of each F_j ($j > K$) are clearly determined by those of F_k by the linear relation (2.4), only by the stoichiometric coefficients σ_{kj} and the reference values at the reactor inlet. The spatial evolution of the system will be then established after the resolution of only the K conservation Eq. (2.3) for the key species, along with the auxiliary Eq. (2.4) to evaluate the rates r_k.

Another way of expressing the relationships given by Eq. (2.4) is by means of the quantities $\mathbb{F}_j$, associated with each component species and defined by:

$$\mathbb{F}_j = F_j - \sum_k \sigma_{kj} F_k; \qquad j > K \tag{2.5a}$$

Thus, it is clear from Eq. (2.4) that each $\mathbb{F}_j$ remains constant and equal to the value at the reactor inlet:

$$\mathbb{F}_j = \mathbb{F}_{j0}; \qquad \mathbb{F}_{j0} = F_{j0} - \sum_k \sigma_{kj} F_{k0}; \qquad j > K \tag{2.5b}$$

The quantities $\mathbb{F}_j$ are said to be *reaction invariant*, since they do not depend on the progress of the chemical reactions. Then, for the local values of each F_j $(j > K)$, instead of Eq. (2.4) it can be alternatively used:

$$F_j = \mathbb{F}_{j0} + \sum_k \sigma_{kj} F_k; \qquad j > K \tag{2.6}$$

The possibility of reaching similar relationships to those established by Eq. (2.4) or (2.6) for different measures of concentration will be explored next.

Use of molar concentrations, $C_j = F_j/q$

Kinetic expressions are frequently expressed in terms of molar concentrations of the species. Thus, the molar concentration can be regarded as a convenient measure of the concentration, especially for didactic purposes.

Considering that the mass flow rate, W, will remain constant for the tubular reactor with plug-flow in steady state (SS), the volumetric flow rate, q, at any axial position is expressed as $q = W/\rho$ or, taking the inlet conditions as a reference, $q = q_0\rho_0/\rho$. Thus, dividing Eq. (2.4) by q:

$$C_j = C_{j0}(\rho/\rho_0) + \sum_k \sigma_{kj}[C_k - C_{k0}(\rho/\rho_0)]; \qquad j > K \tag{2.7a}$$

or alternatively,

$$C_j - \sum_k \sigma_{kj} C_k = (C_{j0} - \sum_k \sigma_{kj} C_{k0})(\rho/\rho_0); \qquad j > K \tag{2.7b}$$

Since the density, ρ, depends in general on all state variables (composition, temperature, T, and pressure, P), Eqs. (2.7a, b) do not allow specifying each C_j for the component species through the knowledge of the C_k only by means of the coefficients

σ_{kj} and the reference values at the inlet. There is a relationship between each C_j and the C_k but involving the variables T and P simultaneously. Therefore, it is concluded that molar concentration does not admit establishing relationships similar to that found for molar flow rates (Eq. 2.4), unless the density can be assumed to remain invariable.

Use of mole fractions: $y_j = F_j/F$

Dividing Eq. (2.4) by the total molar flow rate, F, and taking into account that $y_{j0} = F_{j0}/F_0$:

$$y_j = y_{j0}(F_0/F) + \sum_k \sigma_{kj}[y_k - y_{k0}(F_0/F)]; \qquad j > K \tag{2.8a}$$

or alternatively,

$$y_j - \sum_k \sigma_{kj} y_k = (y_{j0} - \sum_k \sigma_{kj} y_{k0})(F_0/F); \qquad j > K \tag{2.8b}$$

Given the fact that chemical reactions most frequently occur with a change in the number of moles, the case $(F_0/F) \neq 1$ should in general be considered. From Eqs. (2.8a, b), it cannot be directly appreciated if each value of y_j $(j > K)$ is uniquely related to the y_k only by means of the coefficients σ_{kj} and the reference values at the inlet. Then, it is first needed to disclose the dependency of (F_0/F) on the progress of the reactions. For this purpose, $F = \sum_j F_j$ is expressed from Eq. (2.4) for the component species (including the chemically inert species, if any) and the trivial equations $F_k = F_{k0} + (F_k - F_{k0})$ for the key species. The result can be written as:

$$F = F_0 + \sum_k \Delta\sigma_k (F_k - F_{k0}), \tag{2.9}$$

where the summations were swapped, $\sum_j \sum_k \sigma_{kj}(F_k - F_{k0}) = \sum_k (F_k - F_{k0})(\sum_j \sigma_{kj})$, and $\Delta\sigma_k$ is the molar change of the kth normalised canonical reaction:

$$\Delta\sigma_k = \sum_j \sigma_{kj} = 1 + \sum_{j>K} \sigma_{kj} \tag{2.10}$$

After dividing Eq. (2.9) by F and using $y_j = F_j/F$, we get:

$$1 = F_0/F + \sum_k \Delta\sigma_k [y_k - y_{k0}(F_0/F)],$$

from which:

$$F_0/F = \left(1 - \sum\nolimits_k \Delta\sigma_k y_k\right) / \left(1 - \sum\nolimits_k \Delta\sigma_k y_{k0}\right) \tag{2.11}$$

Replacing (F_0/F) from Eq. (2.11) in (2.8b) and after rearranging,

$$\frac{y_j - \sum_k \sigma_{kj} y_k}{1 - \sum_k \Delta\sigma_{kj} y_k} = \frac{y_{j0} - \sum_k \sigma_{kj} y_{k0}}{1 - \sum_k \Delta\sigma_{kj} y_{k0}}; \qquad j > K \tag{2.12}$$

It follows from Eq. (2.12) that the quantities $\mathbb{Y}_j$, defined as:

$$\mathbb{Y}_j = \frac{y_j - \sum_k \sigma_{kj} y_k}{1 - \sum_k \Delta\sigma_{kj} y_k}; \qquad j > K \tag{2.13a}$$

also behave as reaction invariants, $\mathbb{Y}_j = \mathbb{Y}_{j0}$, with:

$$\mathbb{Y}_{j0} = \frac{y_{j0} - \sum_k \sigma_{kj} y_{k0}}{1 - \sum_k \Delta\sigma_{kj} y_{k0}}; \qquad j > K \tag{2.13b}$$

Therefore, given the values of $\mathbb{Y}_{j0}$ at the reactor inlet, each y_j becomes established by:

$$y_j = \mathbb{Y}_{j0} + \sum_k (\sigma_{kj} - \mathbb{Y}_{j0}\Delta\sigma_k) y_k; \qquad j > K \tag{2.13c}$$

We now see that Eq. (2.13c) effectively allows relating the y_j values ($j > K$) linearly with the y_k values only by means of the stoichiometric coefficients σ_{kj} (including their combination $\Delta\sigma_k$) and the reference value at the inlet conditions (y_{j0}).

It is worth noting that by summing the $\mathbb{Y}_j$ as defined in Eq. (2.13a) for $j > K$ the following condition is obtained:

$$\sum_{j>K} \mathbb{Y}_j = 1 \tag{2.14}$$

In addition, for consistency in the formulation, the conservation equations given by Eq. (2.3) for each key species should be expressed in terms of their mole fraction y_k. Taking the derivate of $F_k = y_k F$ with respect to V and considering Eqs. (2.3), (2.9) and (2.11), it is possible to write for each A_k:

$$\left(\frac{1 - \sum_{k'} \Delta\sigma_{k'} y_{k'0}}{1 - \sum_{k'} \Delta\sigma_{k'} y_{k'}}\right) F_0 \frac{dy_k}{dV} = r_k - y_k \left(\sum_{k'} \Delta\sigma_{k'} r_{k'}\right) \tag{2.15}$$

where the terms between brackets are common to all key species.

Equation (2.15) is not attractive for didactic purposes, due to their complexity when compared to, e.g., Eq. (2.3). Thus, it can be concluded that the use of mole fractions will not be an advisable choice for kinetically constrained systems in a CRE course. However, in contexts other than introductory courses, the use of mole fractions can be adequate. In this sense, it should be mentioned that Doherty and co-workers have conveniently employed the reaction invariants $\mathbb{Y}_j$ for application to reactive distillation processes under conditions of chemical equilibrium (see, e.g., [1–3]).

Use of moles per unit mass: $\psi_j = F_j/W$

Since the mass flow rate is uniform in the reactor, the intensive variable defined as $\psi_j = F_j/W$ results in a convenient option (note that it should not be confused with *molality*). Dividing Eq. (2.4) by the total mass flow rate W gives:

$$\psi_j = \psi_{j0} + \sum_k \sigma_{kj}(\psi_k - \psi_{k0}); \qquad j > K \tag{2.16}$$

Equation (2.16) keeps the desired stoichiometric relationships. Moreover, it is also possible to define the quantities Ψ_j:

$$\Psi_j = \psi_j - \sum_k \sigma_{kj}\psi_k; \qquad j > K, \tag{2.17a}$$

which according to Eq. (2.16) are reaction invariants, as they remain constant and equal to the inlet values, regardless of any effect of chemical reactions:

$$\Psi_j = \Psi_{j0}; \qquad \Psi_{j0} = \psi_{j0} - \sum_k \sigma_{kj}\psi_{k0}; \qquad j > K \tag{2.17b}$$

It can, therefore, be used as an alternative to Eq. (2.16):

$$\psi_j = \Psi_{j0} + \sum_k \sigma_{kj}\psi_k; \qquad j > K \tag{2.18}$$

Regarding the conservation equations, with $F_j = W\psi_j$ it can be simply written from Eq. (2.3):

$$W d\psi_k/dV = r_k. \tag{2.19}$$

It should be mentioned that the variable ψ_j has been proposed as a measure of the concentration for certain applications, e.g., by Gillespie and Solomons [4], who define it as *molon*. However, it must be recognised that its occurrence in the literature is very rare and probably unknown to students of an introductory CRE course. Thus, while satisfactory from a practical point of view, the systematic use of such variables in a CRE course may be inappropriate.

Use of mass fractions: $\omega_j = m_j F_j/W$

The mass fractions ω_j differ from the variables ψ_j only by the molar mass of A_j, $\omega_j = m_j\psi_j$. Therefore, it will also be suitable to establish stoichiometric relationships between the quantities of the component species and the key species. From Eq. (2.16):

$$\omega_j = \omega_{j0} + \sum_k (m_j/m_k)\sigma_{kj}(\omega_k - \omega_{k0}); \qquad j > K \tag{2.20}$$

The use of a measure of the concentration based on the mass of each species suggests expressing the stoichiometry and reaction rates in mass terms rather than in molar terms. As such, the net mass production rate of species A_j per unit of time and volume, $\hat{\mathcal{r}}_j$, results:

$$\hat{\mathcal{r}}_j = m_j \mathcal{r}_j \tag{2.21a}$$

Further defining the mass stoichiometric coefficient:

$$\hat{\sigma}_{kj} = \frac{m_j}{m_k}\sigma_{kj}, \tag{2.21b}$$

the stoichiometric relationships (2.2) are rewritten as:

$$\hat{\mathcal{r}}_j = \sum_k \hat{\sigma}_{kj}\hat{\mathcal{r}}_k; \qquad j > K \tag{2.21c}$$

From definition (2.21b), the mass stoichiometric coefficients of the key species in the canonical reactions do not differ from the molar stoichiometric coefficients, i.e., $\hat{\sigma}_{k,k'} = 1$ if $k = k'$ and $\hat{\sigma}_{k,k'} = 0$ otherwise. Moreover, for each canonical reaction, $\mathcal{R}_k$, the summation of the mass stoichiometric coefficients, $\hat{\sigma}_{kj}$, is zero, by virtue of the mass conservation in chemical reactions:

$$1 + \sum_{j>K} \hat{\sigma}_{kj} = 0 \tag{2.21d}$$

Considering the definitions given in Eq. (2.21b), the stoichiometric relationship Eq. (2.20) results in:

$$\omega_j = \omega_{j0} + \sum_k \hat{\sigma}_{kj}(\omega_k - \omega_{k0}); \qquad j > K \tag{2.22}$$

Furthermore, it can be easily verified that the variables Ω_j defined as:

$$\Omega_j = \omega_j - \sum_k \hat{\sigma}_{kj}\omega_k; \qquad j > K \tag{2.23a}$$

also behave as reaction invariants:

$$\Omega_j = \Omega_{j0}; \qquad \Omega_{j0} = \omega_{j0} - \sum_k \hat{\sigma}_{kj}\omega_{k0}; \qquad j > K \tag{2.23b}$$

It is worth noting that if Eqs. (2.23a) are summed over all the component species, including the chemically inert ones, it results in:

$$\sum_{j>K} \Omega_j = 1 \tag{2.23c}$$

which arises by virtue of Eq. (2.21d).

As before, Eq. (2.22) can alternatively be written as:

$$\omega_j = \Omega_{j0} + \sum_k \hat{\sigma}_{kj}\omega_k; \qquad j > K \tag{2.24}$$

Finally, the mass conservation equations for the key species are simply written as:

$$W d\omega_k / dV = \hat{r}_k. \tag{2.25}$$

It can be appreciated that the stoichiometric relationships (Eqs. 2.22 or 2.24) and the mass conservation equations (Eq. 2.25) can be compactly expressed in terms of ω_j. Nevertheless, expressing the stoichiometry of reactions on a mass rather than the usual molar basis is a disadvantage from a didactic point of view. Moreover, since the reaction rates are usually given in terms of molar concentrations, $C_j = \rho\omega_j/m_j$, the molar mass m_j are introduced in the expressions of the $\hat{r}_k$ as well as in the definitions of the $\hat{\sigma}_{kj}$. Such feature can be an inconvenient in the exercises of an introductory CRE course. Conversely, in a real problem, which normally requires the use of more complex models, the above aspects will be irrelevant, as will be discussed in Chap. 3.

2.2.1 Comments on the Usage of Different Measures of the Concentration

It is convenient to remark that the measures of the concentration that allow establishing the stoichiometric relationships in terms of reaction invariants (y_j, ψ_j, and ω_j) for the plug-flow reactor here analysed are easily interchangeable with each other. By replacing $\psi_j = \omega_j/m_j$ and $\psi_k = \omega_k/m_k$ in the definition of Ψ_j (Eq. 2.17b) and considering Eq. (2.21b) and the definition of Ω_j (Eq. 2.23a), it is obtained:

$$\Omega_j = m_j \Psi_j; \qquad j > K \tag{2.26a}$$

In addition, in Appendix 1 the relationships between the variables Ω_j and $\mathbb{Y}_j$ (Eq. 2.13a) is developed, resulting in the reciprocal expressions:

$$\Omega_j = \frac{m_j \mathbb{Y}_j}{\sum\limits_{\beta > K} m_\beta \mathbb{Y}_\beta}; \qquad \mathbb{Y}_j = \frac{\Omega_j / m_j}{\sum\limits_{\beta > K} \Omega_\beta / m_\beta} \tag{2.26b}$$

It follows from Eqs. (2.26a, b) that given one of the sets of variables $\mathbb{Y}_j$, Ω_j or Ψ_j, the others become biunivocally established. Therefore, if for any reaction system one of these set of variables behave as a reaction-invariant set, the other two sets will also behave as such.

Variables like $\mathbb{F}_j$, $\mathbb{Y}_j$, Ω_j or Ψ_j that stoichiometrically relate the amount of an extensive or intensive measure of the component species ($j > K$) with those of the key species are called *component variables.* A formal definition of component variables is given in Chap. 3, where it is also discussed at length that a given component variable may or may be not behave as a reaction invariant, depending on the model and assumptions adopted to describe the reacting systems. For example, recalling the comments on Eq. (2.7a, b), the component variables $\mathbb{C}_j = C_j - \sum_k \sigma_{kj} C_k$ do not behave as reaction invariants for the plug-flow model, unless density is assumed to remain constant.

Finally, it should be emphasised that the definition and explicit use of component variables and reaction invariants can be avoided in an introductory CRE course. The usefulness of such concepts will be explored in Chap. 3 in connection to more complex reaction systems.

2.2.2 Definition of Relative Variables

It has been seen that none of the different measures of concentration explored in the previous section is completely free of inconveniences, from a didactic point of view, for the formulation of conservation balances and the use of stoichiometric relationships in the systems to be analysed in this chapter. Therefore, the use of *relative variables*, defined as the ratio between the extensive variables of the species (such as the molar flow rates, F_j) and some other extensive reference variable, emerges as a trade-off solution. There are several alternatives for such a choice, some of them found in textbooks.

The conversion of a given reactant, usually the limiting one, is one of the most commonly used relative variable for a **single reaction**, when it is probably the most convenient one. If A denotes such a reactant in a continuous-flow reactor, the reference variable is the inlet molar flow rate of A, F_{A0}, and its conversion is $x_A = (F_{A0} - F_A)/F_{A0}$. For the case of **multiple reactions**, the emphasis in the present text, a direct extension of the previous concept should be to choose the inlet molar flow rate of each of the K key species, F_{k0}. However, it is highly unlikely to find a set of K key species having a non-zero value of F_{k0} simultaneously. Even if it were possible to find such a set, the practical significance of the x_k, similar as for the limiting reactant, is also lost. Thus, this alternative will not be explored here. In what follows, two other alternatives will be discussed, possible advantages and disadvantages will be mentioned, and one of them will be selected as the basis for most of the reaction systems in this chapter.

A possible relative variable is the ratio between the number of moles of each species and a reference total number of moles. For continuous-flow reactors, the reference is the total inlet molar flow rate, F_0:

$$\mathscr{y}_j = F_j / F_0 \tag{2.27}$$

Likewise, for a close batch reactor with perfect mixing the reference will be the total number of moles at the beginning of the operation, N_I, and thus $\mathcal{Y}_j = N_j/N_I$.

The definition given in Eq. (2.27) requires only specifying the F_{j0} values at the reactor inlet, resulting in $\mathcal{Y}_{j0} = y_{j0}$, i.e., the inlet mole fraction. Furthermore, from Eq. (2.4), the stoichiometric relationships become:

$$\mathcal{Y}_j = y_{j0} + \sum_k \sigma_{kj}(\mathcal{Y}_k - y_{k0}); \qquad j > K \tag{2.28}$$

whereas, considering Eq. (2.9), the total molar flow rate relative to the total inlet molar flow rate is:

$$\sum_j \mathcal{Y}_j = F/F_0 = \mathcal{Y} = 1 + \sum_k \Delta\sigma_k(\mathcal{Y}_k - y_{k0}) \tag{2.29}$$

The variables $\mathcal{Y}_j$ can be related to the different measures of concentration. Particularly, with respect to mole fractions, $y_j = F_j/F = (F_j/F_0)/(F/F_0)$:

$$y_j = \mathcal{Y}_j/\mathcal{Y} \tag{2.30}$$

The second alternative here considered for a continuous-flow reactor is to take the volumetric inlet flow rate, q_0, as a reference:

$$\breve{C}_j = F_j/q_0 \tag{2.31}$$

which at the inlet of the reactor coincides with the inlet molar concentration, $\breve{C}_{j0} = C_{j0}$.

Similarly, for a close batch reactor with perfect mixing and taking as a reference the total volume at the beginning of the operation, V_I, it will be $\breve{C}_j = N_j/V_I$. Whichever the case, $\breve{C}_j$ and $\mathcal{Y}_j$ are simply related by $\breve{C}_j = y_jC_0$ (or $\breve{C}_j = y_jC_I$)

From Eq. (2.4), the stoichiometric relationships for variables $\breve{C}_j$ result in:

$$\breve{C}_j = C_{j0} + \sum_k \sigma_{kj}\left(\breve{C}_k - C_{k0}\right); \qquad j > K \tag{2.32}$$

The main difference in the use of both variables, $\mathcal{Y}_j$ and $\breve{C}_j$, arises from the information needed to specify the reference conditions, e.g., the inlet conditions for a continuous-flow reactor. The inlet mole fractions $\mathcal{Y}_{j0} = y_{j0}$ are just defined by the inlet composition, a necessary information to undertake any problem. On the other hand, the inlet molar concentrations $\breve{C}_{j0} = C_{j0}$ further require establishing a volumetric variable, either density or total molar concentration: $C_{j0} = y_{j0}C_0 = (\omega_{j0}/m_j)\rho_0$.

Hence, the inlet molar concentrations will depend on temperature and pressure, through C_0 or ρ_0, which can vary according to the stream pre-treatment. Conversely, y_{j0} values are normally expected to remain unchanged. In addition, as will be described later in this chapter (see Sect. 2.6), the volumetric variables in the feed, C_0 or ρ_0, may be superfluous for a continuous-flow reactor with perfect mixing, only the density of the reacting mixture being relevant (and equal to that of the outlet stream). On the other hand, when the reaction kinetics for gas-phase reactions is expressed in terms of partial pressures, there will not be an explicit effect of either the density or the total molar concentration, and only the total pressure needs to be specified (see Sect. 2.5.4).

For the reasons given above, the variables $\mathcal{Y}_j$ will be preferred as a general choice. However, it should be mentioned that the assumption of constant density is frequently explored in introductory CRE courses. Under this condition, $\breve{C}_j = C_j$ and the models used for (ideal) chemical reactors can be formulated only in terms of the molar concentrations of the reactive species.

2.3 Overall Species Balances in Steady-State Flow Systems

Consider a single fluid phase reacting system in steady state, characterised by the number of moles of each species entering and leaving the reactor per unit time, F_{j0} and F_j respectively. An overall steady-state (SS) mass balance for each of the species is simply given by the net difference between F_j and F_{j0} as:

$$F_j - F_{j0} = \mathfrak{R}_j, \tag{2.33}$$

where $\mathfrak{R}_j$ (mol$_j$/s) is the total amount of moles of A_j generated in the entire reactor volume per unit time, or just the *total production rate* of A_j,

$$\mathfrak{R}_j = \int_V r_j dV \tag{2.34}$$

Using the stoichiometric relationships (2.2) ($r_j = \sum_k \sigma_{kj} r_k$), it is possible to write

$$\mathfrak{R}_j = \sum_k \sigma_{kj} \mathfrak{R}_k \tag{2.35}$$

Thus, after replacing the total production rates from Eqs. (2.33) to (2.35):

$$F_j = F_{j0} + \sum_k \sigma_{kj}(F_k - F_{k0}); \qquad j > K \tag{2.36}$$

which is an expression equivalent to that of Eq. (2.4), but with F_j values referred to the reactor outlet.

Process plant balances are one of the instances of direct application of Eq. (2.36). Thus, Eq. (2.36) establish that a given reaction unit will impose $(S - K)$ restrictions to the overall set of unknown variables and that such a unit allows setting up to K flow rates or their differences between the inlet and output. For example, in term of the key species, the conversion of the limiting reactant can be specified, while the remainder differences $(F_k - F_{k0})$ can be set on the basis of assumed selectivities.

Experimental studies are another important application of Eq. (2.36). On the one hand, if the reactions taking place in the system are unknown, Eq. (2.36) can be used to test a possible reaction set by measuring all the inlet and outlet molar flow rates of the experimental reactor. On the other hand, when the occurring reactions are known, Eq. (2.36) shows that, for studying different operating conditions or different catalysts, it will only be necessary to monitor the F_k values of the key species.

Alternatively, the relationships given in Eq. (2.36) can be expressed in terms of the relative variable $\mathcal{Y}_j = F_j/F_0$, according to Eq. (2.28). Similarly, it can also be useful to use the relationships in terms of mole fractions y_j (Eq. 2.13c) if, for example, analytical values of the mole fractions at the inlet and outlet streams are available.

Finally, it should be noted that Eq. (2.36) may be applied for multiple feed or extraction streams, whatever their aggregation state, provided that both F_j and F_{j0} refer to the sum of the inlet and outlet streams, respectively.

2.4 Chemical Equilibrium

Given the inlet composition, the objective is to determine the outlet composition of a steady-state flow homogeneous system, assuming that all the R chemical reactions taking place reached the chemical equilibrium.

To that end, two situations will be considered. The first one corresponds to the case when the outlet temperature and pressure are set. The second one corresponds to the specific case in which only the outlet temperature is specified, but the system evolves at a constant density ρ, which is assumed to be known. That is, either the pair (T, P) or (T, ρ) will be established. Although formulated for open systems, the procedure here presented will also be applicable to closed systems with a given initial composition and stipulating the same conditions at chemical equilibrium.

The thermodynamic equilibrium criterion at a given T and P states that the Gibbs free energy of the system must reach a minimum, which is equivalent to say that the *free energy of reaction* of each linearly independent reaction should be zero. The free energy of reaction for a canonical reaction is $\Delta G_k = \mu_k + \sum_{j>K} \sigma_{kj} \mu_j$, where μ_j is the chemical potential of A_j. Expressing μ_j in terms of the fugacity of each species, f_j, $\mu_j = \mu_j^{ref} + RT\ln(f_j/f_j^{ref})$, the equilibrium condition is then formulated as:

$$\Delta G_k = \Delta G_k^{ref} + RT[\ln(f_k/f_k^{ref}) + \sum_{j>K} \sigma_{kj} \ln(f_j/f_j^{ref})] = 0 \tag{2.37a}$$

$$\Delta G_k^{ref} = \mu_k^{ref} + \sum_{j>K} \sigma_{kj} \mu_j^{ref} \tag{2.37b}$$

where the superscript ref indicates some arbitrary *reference state* at a given temperature, pressure, and composition, and ΔG_k^{ref} is the free energy of reaction in the reference state. We will only consider here two reference states, both defined at the temperate T of the system.

The first one is defined by considering the hypothetical ideal gas behaviour of all the species at $P = 1$ atm, and will be identified by the superscript "$I(1)$". Then, $\Delta G_k^{ref} \equiv \Delta G_k^{I(1)}(T)$, $f_j^{ref} = f_j^{I(1)} = 1$ atm. Each f_j is conveniently expressed through fugacity coefficients, ϕ_j: $f_j = y_j \phi_j P$, with $[P] =$ atm. For an ideal gas, $\phi_j = 1$. This reference state is usually employed for gases or vapours, although it is also used for liquids when an adequate equation of state is available. However, fugacities present a weak dependence on pressure for liquids and thus the coefficients ϕ_j are nearly inversely proportional to P. Hence, even though no theoretical limitation exists for their use for a liquid mixture, another reference state is preferred for such cases, as described in the next paragraph.

The second reference state, the one normally used for liquids, is commonly known as *normal state*, and is defined by considering the species in the pure state at T, P, and in the system phase. It will be identified by the superscript "0", then $\Delta G_k^{ref} \equiv \Delta G_k^0(P, T)$ and $f_j^{ref} = f_j^0$ (for a gas dissolved at temperatures above its critical temperature, a hypothetical condition must be used to evaluate f_j^0). In this case, the f_j are expressed in terms of the activity coefficients γ_j: $f_j = y_j \gamma_j f_j^0$, from which it follows that $\gamma_j = 1$ for a pure component. The ratio $a_j = f_j / f_j^0$ is called the *activity* of species A_j, as defined by Lewis. For a deeper discussion on these topics, see for example, the book by Prausnitz et al. [5].

Whichever the reference state employed, after replacing the expressions of f_j in Eq. (2.37a), the chemical equilibrium conditions can be practically expressed in the form:

$$y_k \prod_{j>K} y_j^{\sigma_{kj}} = K_{y,k}, \tag{2.37c}$$

where $K_{y,k}$ is the so-called *equilibrium constant* (for mole fractions), which collects the remaining terms from Eqs. (2.37a, b). Other measures of concentration can be used to express the equilibrium conditions by considering their relations to mole fractions, as molar concentrations, $C_j = y_j C$ (C: total molar concentration) or, in the case of gas or vapour mixtures, partial pressures, $P_j = y_j P$. These alternatives for the two reference states here considered are shown in Table 2.1, where $\upsilon = 1/C$ is the molar volume in the mixture.

For the actual evaluation of the equilibrium composition, the stoichiometric relationships between the amounts of the reacting species should be considered. Equation (2.13c) should be used if mole fractions y_j are chosen to express the equilibrium composition. However, following the discussion in Sect. 2.2, the extensive variables F_j (for a flow system) and the relative variables $\mathcal{Y}_j$ will be extensively employed in this chapter. Therefore, expressions of chemical equilibrium explicitly accounting for the stoichiometric relationships are displayed in Table 2.2 for the different choices. For the molar flow rates F_j,

Table 2.1 Equilibrium constant expressions for different measures of concentration

Measure of the concentration	Equilibrium constant	Thermodynamic expression	
		Ref.: $\Delta G_k^{I(1)}(T)$	Ref.: $\Delta G_k^0(P, T)$
Mole fraction	$y_k \prod_{j>K} y_j^{\sigma_{kj}} = K_{y,k} =$	$^{(*)}K_k^{I(1)}/(K_{\phi,k} P^{\Delta\sigma_k})$	$K_k^0/K_{\gamma,k}$
Molar concentration	$C_k \prod_{j>K} C_j^{\sigma_{kj}} = K_{C,k} =$	$^{(*)}K_k^{I(1)}/[K_{\phi,k}(P\upsilon)^{\Delta\sigma_k}]$	$K_k^0/(K_{\gamma,k}\upsilon^{\Delta\sigma_k})$
Partial pressure	$^{(*)}P_k \prod_{j>K} P_j^{\sigma_{kj}} = K_{P,k} =$	$K_k^{I(1)}/K_{\phi,k}$	$K_k^0/(K_{\gamma,k} P^{\Delta\sigma_k})$

$$RT\ln K_k^{I(1)} = -\Delta G_k^{I(1)}(T) = -[\mu_k^{I(1)}(T) + \sum_{j>K}\sigma_{kj}\mu_j^{I(1)}(T)]; \quad K_{\phi,k} = \phi_k \prod_{j>K}\phi_j^{\sigma_{kj}}$$

$$RT\ln K_k^0 = -\Delta G_k^0(P, T) = -[\mu_k^0(P, T) + \sum_{j>K}\sigma_{kj}\mu_j^0(P, T)]; \quad K_{\gamma,k} = \gamma_k \prod_{j>K}\gamma_j^{\sigma_{kj}}$$

(*): in these expressions P and P_j should be expressed in atmospheres

Table 2.2 Equilibrium expressions as a function of the constants $K_{y,k}$, using either y_k, F_k or $\mathscr{y}_k$ as the unknowns variables

Expressions for y_j	Chemical equilibrium condition for $K_{y,k}$
$y_j = \mathbb{Y}_{j0} + \sum_k(\sigma_{kj} - \mathbb{Y}_{j0}\Delta\sigma_k)y_k$	$K_{y,k} = y_k \prod_{j>K}\left[\mathbb{Y}_{j0} + \sum_k(\sigma_{kj} - \mathbb{Y}_{j0}\Delta\sigma_k)y_k\right]^{\sigma_{kj}}$
$y_j = \frac{F_j}{F} = \frac{F_{j0}+\sum_k\sigma_{kj}(F_k-F_{k0})}{F_0+\sum_k\Delta\sigma_k(F_k-F_{k0})}$	$K_{y,k} = F_k \frac{\prod_{j>K}\left[F_{j0}+\sum_k\sigma_{kj}(F_k-F_{k0})\right]^{\sigma_{kj}}}{\left[F_0+\sum_k\Delta\sigma_k(F_k-F_{k0})\right]^{\Delta\sigma_k}}$
$y_j = \frac{\mathscr{y}_j}{\mathscr{y}} = \frac{y_{j0}+\sum_k\sigma_{kj}(\mathscr{y}_k-y_{k0})}{1+\sum_k\Delta\sigma_k(\mathscr{y}_k-y_{k0})}$	$K_{y,k} = \mathscr{y}_k \frac{\prod_{j>K}\left[y_{j0}+\sum_k\sigma_{kj}(\mathscr{y}_k-y_{k0})\right]^{\sigma_{kj}}}{\left[1+\sum_k\Delta\sigma_k(\mathscr{y}_k-y_{k0})\right]^{\Delta\sigma_k}}$

$y_j = F_j/F$ was replaced in Eq. (2.37c) and Eqs. (2.4) and (2.9) were used for the component species and the total molar flow F, respectively. Similarly, for the relative variables $\mathscr{y}_j$, Eqs. (2.28)–(2.30) were used. In any case, there are K equations to calculate the key species values y_k, F_k or $\mathscr{y}_k$.

For ideal gases or ideal liquid solutions, $\phi_j = 1$ or $\gamma_j = 1$ respectively, according to the reference state. Then, $K_{\phi,k} = 1$ or $K_{\gamma,k} = 1$ in Table 2.1, whereby the constants $K_{y,k}$ become independent of the equilibrium composition. Otherwise, $K_{\phi,k}$ or $K_{\gamma,k}$ will depend on the unknowns y_k, F_k or $\mathscr{y}_k$, through each y_j.

When the system evolves at a constant density, $\rho = \rho_0$, the volumetric flow rate will also be uniform, $q = q_0$, since the mass flow rate W is conserved. Such an approximation may be reasonable for liquid-phase reactions and even for gases with dilute reactants at constant pressure and temperature. Then, after dividing Eqs. (2.4) and (2.9) by the volumetric flow rate, the stoichiometric relationships and the total molar concentration are written as:

$$C_j = C_{j0} + \sum_k \sigma_{kj}(C_k - C_{k0}); \qquad j > K \tag{2.38}$$

$$C = C_0 + \sum_k \Delta\sigma_k(C_k - C_{k0}) = 1/\upsilon \tag{2.39}$$

Thus, the most practical formulation for the equilibrium condition for constant density will be those in terms of $K_{C,k}$ in Table 2.1, which in terms of the unknowns C_k is explicitly given by:

$$(\rho = \rho_0 \text{ uniform}) \qquad C_k \prod_{j>K} \left[C_{j0} + \sum_k \sigma_{kj}(C_k - C_{k0}) \right]^{\sigma_{kj}} = K_{C,k} \tag{2.40}$$

Specifically for reactions in the liquid phase and the consequent use of the *normal state*, $\Delta G_k^0(P, T)$, it is seen from Table 2.1 that the $K_{C,k}$ values will depend on the molar volume υ if the kth reaction does not conserve the number of moles ($\Delta\sigma_k \neq 0$). Hence, from Eq. (2.39), the $K_{C,k}$ value will depend on the unknowns C_k, even if the solution could be considered as an ideal solution ($K_{\gamma,k} = 1$). The effect of $\upsilon^{\Delta\sigma_k}$ on $K_{C,k}$ cannot be just regarded as negligible and can be particularly significant if the reactants are concentrated, for which C_k or C_{k0} may be of the order of C_0. However, it can be argued that $\Delta\sigma_k \neq 0$ almost invariably implies that the molecular sizes of the species are different. In such a case, the solution should not be expected to behave as ideal, even if the molecules do not differ significantly in their relative attractive forces. In Appendix 2 it is shown that the effect of $\upsilon^{\Delta\sigma_k}$ on $K_{C,k}$, under the constant-density condition, is reasonably compensated by the departure from ideality caused by differences of molecule sizes. However, this does not implies that $K_{\gamma,k}\upsilon^{\Delta\sigma_k} \approx 1$ in the denominator of $K_{C,k}$, as is shown in Appendix 2, where suitable expressions for evaluating $K_{C,k}$ in such a case are also provided.

2.5 Steady-State Tubular Reactors with Plug-Flow Behaviour

The plug-flow hypothesis (PF) applied to a tubular reactor (PFR) implies that temperature and composition are uniform in a given cross section of the reactor and there is no dispersive contribution to the mass and energy transport in the axial direction. For the sake of brevity, we can write the corresponding mass conservation equations from the global balances given in Sect. 2.3. Under the PF hypothesis, Eq. (2.33) is valid at any axial position z in the tube, allowing to express the local value of F_j at a given z, whereas $\mathfrak{R}_j$ are now the total amount of moles of A_j per unit time generated in the volume between $z = 0$ (reactor inlet) and the specific position z considered. Instead of the variable z, we will use hereinafter the reactor volume up to the position z. Then, differentiating Eqs. (2.33) and (2.34) with respect to V:

2.5 Steady-State Tubular Reactors with Plug-Flow Behaviour

$$dF_j/dV = r_j; \quad F_j(0) = F_{j0} \quad (2.41)$$

where r_j is to be evaluated at the local temperature and composition at z. Particularly for the key species:

$$dF_k/dV = r_k; \quad F_k(0) = F_{k0} \quad (2.42a)$$

The stoichiometric relationships in Eq. (2.36) are also valid:

$$F_j = F_{j0} + \sum_k \sigma_{kj}(F_k - F_{k0}); \quad j > K \quad (2.42b)$$

which are now applied locally at a generic position z.

Equations (2.41) and (2.42a, b) can be re-written in terms of the relative variables $\mathcal{Y}_j = F_j/F_0$ (Eq. 2.27), just by dividing them by F_0. Then, it results:

$$F_0 \, d\mathcal{Y}_k/dV = r_k; \quad \mathcal{Y}_k(0) = y_{k0} \quad (2.43a)$$

$$\mathcal{Y}_j = y_{j0} + \sum_k \sigma_{kj}(\mathcal{Y}_k - y_{k0}); \quad j > K \quad (2.43b)$$

To express the molar concentrations $C_j = y_j C$ on which the reaction rates are supposed to depend, it is replaced $y_j = \mathcal{Y}_j/\mathcal{Y}$ (Eq. 2.30):

$$C_j = (\mathcal{Y}_j/\mathcal{Y})C \quad (2.44)$$

The ratio $\mathcal{Y} = (F/F_0)$ given in Eq. (2.29) is repeated here for the sake of clarity:

$$\mathcal{Y} = 1 + \sum_k \Delta\sigma_k(\mathcal{Y}_k - y_{k0}) \quad (2.45a)$$

Then,

$$C_j = y_j C = \frac{\mathcal{Y}_j}{1 + \sum_k \Delta\sigma_k(\mathcal{Y}_k - y_{k0})} C \quad (2.45b)$$

where $\mathcal{Y}_j = \mathcal{Y}_k$ for the key species, while for the component species $\mathcal{Y}_j$ is expressed by Eq. (2.43b).

For solving the mass conservation Eqs. (2.43a), the dependence of the total molar concentration C on the state variables, composition (e.g., mole fractions y_j), temperature T (for non-isothermal reactors) and pressure P (of less relevance, due to the usually small variations along the reactor) must be established. Two possible scenarios can be visualised to this end.

On the one hand, when a rigorous evaluation of C is required, an equation of state allowing to express, for example, the molar volume $v = 1/C$ as a function of y_j, T and P, can be used. This possibility will be more usual for gases and vapour mixtures. Although clearly applicable also for liquid-phase reactions, a simpler and probably accurate enough approach in this case arises from writing $\upsilon = \sum_j y_j \upsilon_j$, with partial molar volumes υ_j estimated at inlet conditions ($\upsilon_j = \upsilon_{j0}$), considering their relatively weak dependence on composition and temperature.

On the other hand, certain simplifications for evaluating C may be satisfactory in some practical cases and are definitely valuable for didactic purposes. These simplifications will be explored in the following subsections.

In certain occasions it can be convenient to express C_j in terms of $\mathcal{Y}_j$ and ρ instead of $\mathcal{Y}_j$ and C. In principle, this is achieved by taking $C = \rho/\overline{m}$ in Eq. (2.44), where $\overline{m} = \sum_j m_j y_j = (\sum_j m_j \mathcal{Y}_j)/\mathcal{Y}$ is the mean molar mass. However, a more conceptual expression can be obtained by writing for the molar masses of the key species, $m_k = -\sum_{j>K} \sigma_{kj} m_j$, as follows from mass conservation in the chemical reactions. Besides, considering Eq. (2.43b) for the component species, the following relations can be proved to hold: $\sum_j m_j \mathcal{Y}_j = \sum_{j>K} m_j (y_{j0} - \sum_k \sigma_{kj} y_{k0}) = \overline{m}_0$ (i.e., the mean molar mass at the inlet); $\overline{m} = \overline{m}_0/\mathcal{Y}$. Thus, after replacing $C = \rho/\overline{m} = \mathcal{Y}\rho/\overline{m}_0$ in Eq. (2.44):

$$C_j = \mathcal{Y}_j \rho/\overline{m}_0 \tag{2.45c}$$

2.5.1 The Expansion Factor

For didactic purposes, two simplifications are normally used: the assumption of constant density, usually for liquids or diluted gases, and ideal gas behaviour when the fluid phase is a gas. The link between these two simplifications can be conveniently discussed by considering the *expansion factor*:

$$\varepsilon = \rho_0/\rho \tag{2.46}$$

which expresses the ratio between the volume occupied by one kg of mixture at a given position in the reactor and the volume occupied at the inlet. Since the mass flow rate W remains constant, expressing $W = \rho q = \rho_0 q_0$ gives the following equivalence in terms of the volumetric flow rates, q and q_0:

$$\varepsilon = q/q_0 \tag{2.47}$$

Furthermore, expressing $q = F/C$ and $q_0 = F_0/C_0$, Eq. (2.47) can be re-written as:

$$\varepsilon = (F/F_0)(C_0/C) \tag{2.48}$$

And finally, keeping in mind the definition $\mathcal{Y} = (F/F_0)$ and the expression given in Eq. (2.45a), it is obtained:

$$\varepsilon = \left[1 + \sum_k \Delta\sigma_k(\mathcal{Y}_k - y_{k0})\right](C_0/C) \tag{2.49}$$

Equation (2.49) discriminates the expansion of the mixture due to the change in the total number of moles generated by the reactions and the change caused by the modification of the total molar concentration, which is in general triggered by variations in the state variables. The former is quantified by the factor $\mathcal{Y} = F/F_0 = [1 + \sum_k \Delta\sigma_k(\mathcal{Y}_k - y_{k0})]$, while the latter is quantified by the ratio C_0/C.

To express the molar concentration C_j in terms of ε, $\mathcal{Y} = (F/F_0)$ from Eq. (2.48) is replaced in Eq. (2.44):

$$C_j = \mathcal{Y}_j C_0/\varepsilon \tag{2.50}$$

Thus, when assuming constant density (Sect. 2.5.3), C_j will be evaluated from Eq. (2.50) directly with $\varepsilon = 1$. On the other hand, when the fluid phase is a gas, Eq. (2.50) will be used with ε from Eq. (2.49), and $C_0/C = (P_0/P)(T/T_0)$ for the ideal gas assumption.

2.5.2 The Residence Time

As the plug-flow hypothesis neglects the dispersion, all species will have the same time to react from the reactor inlet up to a given position inside the reactor. Such a time is called *residence time* and is not only conceptually relevant from a kinetic point of view, but also because it facilitates the comparison with other systems, such as batch reactors.

When the fluid stream advances a differential distance dz with linear velocity u, the elapsed time is $d\theta = dz/u$, or equivalently $d\theta = dV/q$, which we may write in the form:

$$dV/d\theta = \varepsilon q_0 \tag{2.51}$$

After replacing dV from Eq. (2.43a) in Eq. (2.51) and using $C_0 = F_0/q_0$:

$$C_0 \, d\mathcal{Y}_k/d\theta = \varepsilon r_k; \qquad \mathcal{Y}_k(0) = y_{k0} \tag{2.52}$$

Equation (2.52) shows that the evolution of the composition, expressed by the set of relative variables $\{\mathcal{Y}_k\}$, will depend on the residence time θ needed by the stream to reach a given position in the reactor. Notwithstanding the undoubtedly conceptual significance of θ, it is not a practical independent variable as it depends on the density change of the mixture. For example, if the reactor volume V is specified, the residence time at the reactor outlet, θ_s, is not. Hence, in such a case Eq. (2.52) must be solved simultaneously with Eq. (2.51), considering the dependence of ε on the $\mathcal{Y}_k$ and (C/C_0) (see Eq. 2.49). If the value of θ_s is effectively required, it can be obtained by proceeding in such a way.

Alternatively, the conservation equations in terms of V (Eqs. 2.43a, b) can be solved in a first step and the value of θ_s evaluated a posteriori by integration of Eq. (2.51):

$$\theta_s = q_0 \int_V \frac{1}{\varepsilon} dV \tag{2.53}$$

To employ Eq. (2.53), $\varepsilon(V)$ should be locally evaluated from Eq. (2.49).

2.5.3 Uniform Density

The case of constant density is usually explored in introductory CRE courses, an assumption reasonably acceptable for liquid mixtures in particular or diluted reactants in general. Then, Eq. (2.50) with $\varepsilon = 1$ becomes:

$$(\varepsilon = 1) \qquad C_j = y_j C_0 \tag{2.54}$$

Thus, Eqs. (2.43a, b) can be solved together with the expressions (2.54), with the evident simplification introduced by Eq. (2.54) instead of the general Eq. (2.50).

However, the usual way to proceed when $\varepsilon = 1$ is to employ the mass balances directly in terms of molar concentrations. By using Eq. (2.54) in Eqs. (2.43a, b), and taking into account that $F_0/C_0 = q_0$:

$$(\varepsilon = 1) \qquad q_0 dC_k/dV = r_k; \qquad C_k(0) = C_{k0} \tag{2.55a}$$

$$(\varepsilon = 1) \qquad C_j = C_{j0} + \sum_k \sigma_{kj}(C_k - C_{k0}); \qquad j > K \tag{2.55b}$$

The production rates r_k in Eq. (2.55a) retain their (assumed) intrinsic dependence on C_j. Additionally, Eq. (2.51) with $\varepsilon = 1$ indicates the proportionality between V and the residence time, $\theta = V/q_0$, so that Eq. (2.55a) can alternatively be written as:

$$dC_k/d\theta = r_k; \qquad C_k(0) = C_{k0} \tag{2.55c}$$

The whole formulation is therefore stated in terms of volumetric properties (C_j, C_{j0}, V/q_0).

2.5.4 Ideal Gas Behaviour

The ideal gas assumption will be applicable, at least as a reasonable approximation, for a significant number of practical cases. This allows the use of:

$$(\text{ideal gas}) \qquad C = P/(RT) \tag{2.56a}$$

If the PFR has been adequately designed, the pressure drop should be moderate. In addition, if the temperature variations are also moderate, it can be seen that the process will essentially occur at constant total molar concentration. This condition does not imply constant density, since ρ will vary if any of the reactions involve a change in the number of moles, according to Eq. (2.49).

The equation balances to be used are then Eq. (2.43a, b) with molar concentrations evaluated from Eq. (2.45b), which using Eq. (2.56a) results in:

$$\text{(ideal gas)} \qquad C_j = \frac{\mathscr{y}_j}{1 + \sum_k \Delta\sigma_k(\mathscr{y}_k - y_{k0})} P/(RT) \tag{2.56b}$$

where the variations of T and P, if significant, will arise from the energy and momentum conservation equations.

It is worth mentioning that if some departure from the ideal gas behaviour is to be considered for a more rigorous evaluation of C, the compressibility factor, $Z = P/(RTC)$, can be assumed to be constant for moderate changes of T and P, provided that the relative attractive/repulsive forces between the different species do not strongly differ from each other. Then, evaluating Z at the inlet conditions ($Z = Z_0$), $C = P/(RTZ_0)$ can be used instead of Eq. (2.56a), and from Eq. (2.45b):

$$(Z = Z_0) \qquad C_j = \frac{\mathscr{y}_j}{1 + \sum_k \Delta\sigma_k(\mathscr{y}_k - y_{k0})} P/(RTZ_0) \tag{2.56c}$$

In addition, it must be considered that the kinetic expressions for gas phase reactions are frequently expressed in terms of partial pressures, $P_j = y_j P$, instead of molar concentrations. Using y_j from Eq. (2.45b):

$$P_j = \frac{\mathscr{y}_j}{1 + \sum_k \Delta\sigma_k(\mathscr{y}_k - y_{k0})} P \tag{2.57}$$

It follows from the equation balances (2.43a, b) and from Eq. (2.57), that no volumetric property will be required to determine the reactor behaviour and it will be unnecessary to specify the actual behaviour of the gas phase.

It was mentioned in Sect. 2.2.2 the possibility of using the relative variable $\breve{C}_j = F_j/q_0$ (Eq. 2.31) rather than the $\mathscr{y}_j$. It is clear that the definition of $\breve{C}_j$ requires the specification of the volumetric flow rate q_0. However, as was discussed just above, no volumetric property is needed when the reaction rates are given in terms of partial pressures and therefore the value q_0 will be superfluous. This fact is masked when the mass balances are expressed in terms of the variables $\breve{C}_j$.

2.5.5 Temperature–Composition Relationship for Adiabatic Operations

Plug-flow reactors under steady-state adiabatic operation allow the temperature variations to be related to those of composition. Without detailing the derivation, the following thermal energy balance results:

$$\hat{c}_p W(dT/dV) = -\sum_k \Delta H_k \mathscr{r}_k - q_v; \qquad T(0) = T_0 \tag{2.58a}$$

where q_v is the rate of heat transferred to an external fluid per unit volume of reactor. Equation (2.58a) strictly assumes uniform pressure and neglects heat generation by viscous dissipation effects, which are justifiable assumptions in most practical cases. $\hat{c}_p$ and $\Delta H_k = h_k + \sum_{j>K} \sigma_{kj} h_j$ are the local values of the specific heat capacity of the mixture and the reaction enthalpy of the kth canonical reaction, respectively. In turn, h_j is the partial molar enthalpy of species A_j.

For the adiabatic operation, $q_v = 0$, and after replacing the $\mathscr{r}_k$ from Eq. (2.43a):

$$\hat{c}_p \overline{m}_0 dT = -\sum_k \Delta H_k d\mathscr{y}_k \tag{2.58b}$$

where $\overline{m}_0 = W/F_0$ was considered for the mean molar mass at the reactor inlet.

In order to obtain a linear relationship between temperature and composition, both $\hat{c}_p$ and ΔH_k must be considered as constants. To this end, the values at a certain reference condition may be used. Taking the value of $\hat{c}_p$ at the inlet conditions as a reference ($\hat{c}_p \equiv \hat{c}_{p0}$), it is defined:

$$\tilde{c}_{p0} = \hat{c}_{p0} \overline{m}_0, \tag{2.59}$$

where $\tilde{c}_{p0}$ is the molar heat capacity in the feed.

Replacing Eq. (2.59) in (2.58b):

$$\tilde{c}_{p0} dT = -\sum_k \Delta H_k d\mathscr{y}_k \tag{2.60}$$

Upon integration of Eq. (2.60), assuming ΔH_k as constant, but without specifying the reference conditions for its evaluation (which could also be the inlet conditions):

$$T - T_0 = \sum_k (-\Delta H_k / \tilde{c}_{p0})(\mathscr{y}_k - y_{k0}) \tag{2.61}$$

2.6 Steady-State Flow Reactors with Perfect Mixing Behaviour (CSTRs)

The mass conservation equation in a continuous-flow stirred tank reactor (CSTR), in steady state and under the assumption of perfect mixing (i.e., the fluid contained in the tank presents uniform properties), will be here addressed. Although the expressions to be considered do not distinguish whether the fluid inside the tank is a liquid or a gas mixture, any specific reference will be made to liquid mixtures, as CSTRs are used in practice for liquid-phase reactions.

It will be taken into account the possibility of the CSTR having multiple independent feed streams, e.g., liquid mixtures with different reactants, or even solid, vapour or gas streams. It is assumed that the agitation is vigorous enough to allow the complete and (essentially) instantaneous mixing of the feed streams with the tank content, whatever the phase of the streams is. Thus, when referring to the inlet molar flow rates, F_{j0} and F_0, the contribution of all the feed streams will be considered, i.e., the sum of the moles of A_j and the sum of total moles entering the reactor, respectively. In the same way, W represents the total mass flow rate from all the feeding streams. Having made these clarifications, we will simply refer to "reactor inlet", unless any specific identification is due.

On the contrary, the following formulation is restricted to a single liquid extraction stream from the tank, with the same properties as those of the tank content (C_j, T) and, for the SS under consideration, with the mass flow rate W equal to that of the feed.

For establishing the mass balances per species, the control volume is the volume occupied by the fluid inside the tank. We may use again Eqs. (2.33) and (2.34) for each key species. Upon consideration of perfect mixing behaviour in Eq. (2.34), $\mathfrak{R}_k = V\mathscr{r}_k$, and then:

$$F_k - F_{k0} = V\mathscr{r}_k \tag{2.62}$$

where the production rates $\mathscr{r}_k$ depend on the uniform values of C_j and T in the reacting mixture inside the tank. The stoichiometric relationships given by Eq. (2.36) remain unchanged, but are repeated here for completeness:

$$F_j = F_{j0} + \sum_k \sigma_{kj}(F_k - F_{k0}); \qquad j > K \tag{2.63}$$

Alternatively, using the relative variable $\mathscr{Y}_j = F_j/F_0$, Eqs. (2.62) and (2.63) are easily re-written as:

$$F_0(\mathscr{Y}_k - \mathscr{Y}_{k0}) = V\mathscr{r}_k \tag{2.64a}$$

$$\mathscr{Y}_j = \mathscr{Y}_{j0} + \sum_k \sigma_{kj}(\mathscr{Y}_k - \mathscr{Y}_{k0}); \qquad j > K \tag{2.64b}$$

where $\mathcal{Y}_{j0} = F_{j0}/F_0$ will only coincide with the true molar fraction, i.e., $\mathcal{Y}_{j0} = y_{j0}$, if there is a single feed stream. To express the molar concentration C_j inside the reactor, the same relationships in Eq. (2.45b) are valid, but replacing y_{k0} by $\mathcal{Y}_{k0} = F_{k0}/F_0$:

$$C_j = \frac{\mathcal{Y}_j}{1 + \sum_k \Delta\sigma_k(\mathcal{Y}_k - \mathcal{Y}_{k0})} C \tag{2.65}$$

To complete the formulation, the total molar concentration C (or equivalently, $\rho = \overline{m}C$ or $\upsilon = 1/C$) must be estimated, for which the guidelines given in Sect. 2.5 can be consulted.

It should be emphasised that C is the only volumetric property required and it corresponds to the outlet conditions. Conversely, the volumetric properties at the reactor inlet are irrelevant. However, a frequent approach in CRE courses for the study of CSTRs is to consider a liquid feed stream presenting the same density, ρ_0, as that in the tank, ρ. Hence, the inlet and outlet volumetric flow rates are equal, $q_0 = q$, in which case $\mathcal{Y}_j = C_j/C_0$ and Eqs. (2.64a, b) can be written as follows:

$$q_0(C_k - C_{k0}) = V r_k \tag{2.66a}$$

$$C_j = C_{j0} + \sum_k \sigma_{kj}(C_k - C_{k0}); \qquad j > K \tag{2.66b}$$

As for the PFR, a formulation in terms of volumetric properties $(C_j,\ C_{j0},\ V/q_0)$, where $\theta = V/q_0$ is the average residence time in the tank, is obtained.

Equations (2.66a, b) can be used in a general case, provided that the density ρ of the liquid mixture inside the tank is the property employed to define all the volumetric variables. In such a case, the values C_{j0} and $q_0 = q$ should not be confused with the actual physical values in the feed stream, but to hypothetical values in a single feed stream with density $\rho_0 = \rho$.

2.6.1 Temperature–Composition Relationships for Adiabatic Operation

Under certain simplifications, the operating temperature can be linearly related to the composition inside the tank for an adiabatic operation. A general thermal energy balance for stirred tank reactors with perfect mixing is presented in Sect. 2.8 (Eq. 2.92), which for the present situation at SS and a single outlet stream is written as follows:

$$\sum_\alpha W_\alpha[\hat{c}_{p,\alpha}(T_\alpha - T) - \Delta\hat{H}_\alpha^d] = V\sum_k r_k(\Delta H_k) + Q^* \tag{2.67}$$

where Q^* is the rate of heat transferred from the tank to a heat-exchange device, ΔH_k is the reaction enthalpy of the kth canonical reaction evaluated at the temperature T and composition in the tank. The subscript "α" refers to the different feed streams, $\hat{c}_{p,\alpha}$ is the specific heat capacity averaged between T_α and T (note that no specific symbol is employed to denote such an average).

$\Delta\hat{H}_\alpha^d$ in Eq. (2.67) is the specific heat of mixing of the feed stream α upon dissolution in the tank solution at the temperature T, which can be written in terms of the specific heat of mixing of each species A_j at T, $\Delta\hat{h}_{j\alpha}^d$ as follows:

$$\Delta\hat{H}_\alpha^d = \sum_j \omega_{j\alpha}\Delta\hat{h}_{j\alpha}^d \tag{2.68}$$

where $\omega_{j\alpha}$ is the mass fraction of A_j in the feed stream α.

The total mass flow rate fed to and extracted from the reactor is:

$$W = \sum_\alpha W_\alpha \tag{2.69}$$

Then, the following average values of the variables at the "reactor inlet" can be defined:

$$T_0 = \left(\sum_\alpha W_\alpha \hat{c}_{p,\alpha} T_\alpha\right) / (W\hat{c}_{p0}) \tag{2.70a}$$

$$\hat{c}_{p0} = \left(\sum_\alpha W_\alpha \hat{c}_{p,\alpha}\right) / W \tag{2.70b}$$

$$\Delta\hat{H}_0^d = \left(\sum_\alpha W_\alpha \Delta\hat{H}_\alpha^d\right) / W \tag{2.70c}$$

After replacing Eq. (2.70a–c) in Eq. (2.67):

$$\hat{c}_{p0}W(T_0 - T) = W\Delta\hat{H}_0^d + V\sum_k r_k(\Delta H_k) + Q^* \tag{2.71a}$$

Equation (2.71a) can be re-written in a molar basis with the following substitution:

$$\hat{c}_{p0}W = \tilde{c}_{p0}F_0; \qquad \Delta\hat{H}_0^d W = \Delta H_0^d F_0$$

where $\tilde{c}_{p0}$ and ΔH_0^d are the heat capacity and heat of mixing per total mole of the feed. Then,

$$\tilde{c}_{p0}F_0(T_0 - T) = F_0\Delta H_0^d + V\sum_k r_k(\Delta H_k) + Q^* \tag{2.71b}$$

Finally, replacing the values of r_k from Eq. (2.64a) and considering $Q^* = 0$ for an adiabatic operation:

$$T = T_0 + (-\Delta H_0^d)/\tilde{c}_{p0} + \sum_k [(-\Delta H_k)/\tilde{c}_{p0}](y_k - y_{k0}) \tag{2.72}$$

A frequently made assumption in introductory CRE courses is to consider $\tilde{c}_{p0}$ and ΔH_k as independent of the temperature, and ΔH_0^d as negligible. This allows establishing a linear relationship between temperature and composition in the reactor (and hence at the outlet), which in turn facilitates the calculation of the state variables in the reactor. The first assumption will often be reasonable, although it requires that a relatively low difference $(T - T_0)$ results. On the other hand, ignoring ΔH_0^d will be acceptable for a great variety of cases, albeit it will be inappropriate in case of significant heats of solvation, as, e.g., for the dissolution of a strong acid or base in an aqueous solution.

2.7 Closed Reactors with Perfect Mixing (Batch Reactors)

The operation of a closed stirred tank reactor, or simply "*batch reactor*" (BR), will be analysed here. A liquid-phase mixture will be considered, as is always involved in the practical applications of this type of reactor. It will be also assumed that the agitation is intense enough to maintain uniform composition and temperature all over the reactor content. The mass conservation equation for each species is:

$$dN_j/dt = V r_j; \qquad N_j(0) = N_{jI} \tag{2.73}$$

where N_j is the total number of moles of A_j, V is the instantaneous volume of the mixture and r_j is the production rate of A_j per unit volume, evaluated at the instantaneous temperature and composition in the reactor.

Since the total mass is conserved in time, $M = \rho V = \rho_I V_I$, where the subscript "I" refers to the initial operation time, $t = 0$ for convenience. Defining the expansion factor:

$$\varepsilon = \rho_I/\rho = V/V_I, \tag{2.74}$$

we may re-write Eq. (2.73) as:

$$dN_j/dt = \varepsilon V_I r_j; \qquad N_j(0) = N_{jI} \tag{2.75}$$

Specifically for the key species:

$$dN_k/dt = \varepsilon V_I r_k; \qquad N_k(0) = N_{kI} \tag{2.76a}$$

The stoichiometric relationships between production rates, $r_j = \sum_k \sigma_{kj} r_k$, allow establishing relationships between the number of moles, by replacing the values of r_j

of the component species and the r_k for the key species from Eqs. (2.75) to (2.76a), and subsequent integration from the initial conditions:

$$N_j = N_{jI} + \sum_k \sigma_{kj}(N_k - N_{kI}); \quad j > K \tag{2.76b}$$

The relative variable $\mathcal{Y}_j$ is defined as $\mathcal{Y}_j = N_j/N_I$, where N_I is the total number of moles at $t = 0$. In terms of $\mathcal{Y}_j$, Eqs. (2.76a, b) results in:

$$C_I \, d\mathcal{Y}_k/dt = \varepsilon r_k; \quad \mathcal{Y}_k(0) = y_{kI} \tag{2.77a}$$

$$\mathcal{Y}_j = y_{jI} + \sum_k \sigma_{kj}(\mathcal{Y}_k - y_{kI}); \quad j > K \tag{2.77b}$$

where $C_I = N_I/V_I$ is the total molar concentration at $t = 0$, and it was considered that $\mathcal{Y}_{jI} = N_{jI}/N_I = y_{jI}$.

In addition, if N is the total number of moles at a given time t, the total molar concentration is simply $C = N/V$, from which the expansion factor ε (Eq. 2.74) can be expressed as:

$$\varepsilon = (N/N_I)(C_I/C) \tag{2.78}$$

The ratio $\mathcal{Y} = N/N_I$ arises by adding $\mathcal{Y}_j = N_j/N_I$ for all species, with $\mathcal{Y}_k = y_{kI} + (\mathcal{Y}_k - y_{kI})$ for the key species and Eq. (2.77b) for the component species:

$$N/N_I = \mathcal{Y} = 1 + \sum_k \Delta\sigma_k(\mathcal{Y}_k - y_{kI}) \tag{2.79}$$

leading to

$$\varepsilon = \left[1 + \sum_k \Delta\sigma_k(\mathcal{Y}_k - y_{kI})\right](C_I/C) \tag{2.80}$$

The molar concentration of each species, on which the reaction rates are supposed to depend, can then be evaluated from:

$$C_j = N_j/(\varepsilon V_I) = \mathcal{Y}_j C_I/\varepsilon \tag{2.81}$$

By replacing ε from Eq. (2.80):

$$C_j = \frac{\mathcal{Y}_j}{1 + \sum_k \Delta\sigma_k(\mathcal{Y}_k - y_{kI})} C, \tag{2.82}$$

which is employed with $\mathcal{Y}_j = \mathcal{Y}_k$ for the key species and $\mathcal{Y}_j$ from Eq. (2.77b) for the component species.

Finally, if the density is assumed to remain constant throughout the entire operation, $\varepsilon = 1$, in which case Eq. (2.81) results in $C_j = \mathcal{Y}_j C_I$, and the mass balances given by Eq. (2.77a, b) can be re-written in the form frequently found in textbooks:

$$dC_k/dt = r_k; \qquad C_k(0) = C_{kI} \tag{2.83a}$$

$$C_j = C_{jI} + \sum_k \sigma_{kj}(C_k - C_{kI}); \qquad j > K \tag{2.83b}$$

A comparison between the BR and PFR for liquid-phase reactions at the same levels of constant pressure and temperature and the same initial composition ($y_{jI} = y_{j0}$) can be carried out. Considering that under such conditions $C_I = C_0$ will hold, from comparing the pairs of expression (2.52, 2.77a) and (2.49, 2.80) the same $\mathcal{Y}_j$ values will result for the same value of reaction time t in the BR and of residence time θ in the PFR, regardless of possible density changes. The condition about pressure can be relaxed, due to the very low liquid compressibility, under moderate differences in both systems. As discussed in the next section, adiabatic operations will also keep the similarity between the BR and PFR.

Although a gas-phase operation in a BR is not common, it can be compared to the operation in a PFR. In a BR, the volume and hence the gas density will remain constant, while in a PFR the density will usually change, as discussed in Sect. 2.5. Then, the similarity will only be maintained at conditions leading to keep constant density in the PFR (strictly, $\Delta\sigma_k = 0$, and uniform temperature and pressure). In this case ($\varepsilon = 1$), the similarity follows from comparing the pairs of expressions (2.55c, 2.83a) and (2.55b, 2.83b).

2.7.1 Temperature–Composition Relationships for Adiabatic Operation

For the case of an adiabatic operation in a BR, temperature variations can also be related to those of composition. From the general thermal energy balance for stirred tank reactors with perfect mixing presented in Sect. 2.8 (Eq. 2.92) applied to a BR:

$$M\hat{c}_p(dT/dt) = V\sum_k(-\Delta H_k)r_k - Q^*; \qquad T(0) = T_I, \tag{2.84}$$

where $\hat{c}_p$ and ΔH_k are the values of the specific heat capacity and of the reaction enthalpy of the kth canonical reaction and Q^* is the rate of heat transferred from the tank to a heat-exchange device.

For an adiabatic operation, $Q^* = 0$, and replacing the values of r_k from the mass conservation Eq. (2.77a) and taking into account that $\varepsilon = V/V_I$ (Eq. 2.74):

$$\hat{c}_p \overline{m}_I dT = \sum_k (-\Delta H_k) d\mathcal{Y}_k, \tag{2.85}$$

where $\overline{m}_I = M/N_I$ is the initial mean molar mass.

In a similar way as in Sects. 2.5.5 and 2.6.1, a linear relationship between temperature T and composition $\mathcal{Y}_k$ can be obtained by assuming that $\hat{c}_p$ is evaluated at the initial conditions (i.e., $\hat{c}_p \equiv \hat{c}_{pI}$) and that the ΔH_k are evaluated at a certain reference temperature. Using the initial molar heat capacity, $\tilde{c}_{pI} = \hat{c}_{pI}\overline{m}_I$, it is obtained from Eq. (2.85):

$$T - T_I = \sum_k (-\Delta H_k/\tilde{c}_{pI})(\mathcal{Y}_k - y_{kI}) \tag{2.86}$$

Finally, comparing Eqs. (2.85) and (2.58b) (or alternatively, Eqs. 2.86 and 2.61), it can be observed that, for liquid-phase operation, the equivalence between a BR and a PFR discussed for an isothermal operation will also be valid for an adiabatic operation, provided that $T_I = T_0$.

2.8 Open Reactors with Perfect Mixing in Transient Regime

Operation of stirred tank reactors in batch mode and with the continuous addition of a stream during the reaction time is frequently found in practice for a variety of reasons. As typical examples: low solubility of a reactant (e.g., a gas species), to improve selectivity (e.g., by maintaining a low concentration of a specific species), for safety purposes (e.g., to moderate the release of heat from fast and highly exothermic reaction). The extraction of some product may also be practised for equilibrium-limited reactions. This kind of systems are referred to as semi-continuous or "*semi-batch reactors*" (SBR).

Furthermore, the start-up of CSTRs will follow a transient evolution until reaching the corresponding steady state, which admits a similar description as for a SBR.

As in Sects. 2.6 and 2.7, we will consider liquid-phase reactions and that the perfect mixing hypothesis can be assumed. In addition, any vapour or gas present in the reactor free-space above the liquid level is ignored, because of the negligible mass compared to the mass of the liquid mixture.

Total mass balance:

$$\frac{dM}{dt} = (\sum_\alpha W_\alpha) - W; \qquad M(0) = M_I, \tag{2.87}$$

where M is the instantaneous liquid mass in the reactor and M_I is the initial value ($t = 0$), W is the mass flow rate of a stream eventually extracted at the same conditions of the liquid mixture in the reactor, which will be referred to as the *homogeneous stream*. Any other different stream is identified with the subscript "α" and their mass flow rates W_α are added in the second term of Eq. (2.87), where W_α is positive for inlet streams (solid, liquid, or gaseous) and negative for outlet streams (e.g., vapour extraction in equilibrium with the liquid mixture or an extraction through a membrane).

Any of the mentioned streams may have a time-varying mass flow rate, either pre-set or dependent on the evolution of the system. Its composition could be constant, usually for inlet streams, or time-varying for, e.g., a vapour extraction in equilibrium with the liquid mixture in the tank.

In terms of volumetric variables, with $M = V\rho$, and $W = \rho q$:

$$\frac{d(V\rho)}{dt} = \sum_\alpha W_\alpha - \rho q; \qquad V(0) = V_I \tag{2.88}$$

Mass balance for each species:

$$\frac{dN_j}{dt} = \sum_\alpha F_{j\alpha} - F_j + V\mathscr{r}_j; \qquad N_j(0) = N_{jI} \tag{2.89}$$

where $F_{j\alpha}$ is the molar flow rate of A_j in the inlet/outlet stream α, and the molar flow rate F_j corresponds to the homogeneous stream. The latter can be expressed as:

$$F_j = qC_j = qN_j/V = N_j/\theta \tag{2.90}$$

where the ratio $\theta = V/q = M/W$ will generically be time dependent. Equation (2.89) then results in:

$$\frac{dN_j}{dt} = \sum_\alpha F_{j\alpha} - N_j/\theta + V\mathscr{r}_j; \qquad N_j(0) = N_{jI} \tag{2.91}$$

The conservation Eq. (2.91) will be used directly in terms of the number of moles, N_j, of each species A_j. Albeit it is possible to use a relative variable, the open and simultaneous transient nature of the system precludes the choice of a suitable reference quantity. For example, for a batch reactor (Sect. 2.7), the relative variable $\mathscr{y}_j = N_j/N_I$ was shown to be useful. This happens because the reference quantity N_I is proportional to the total mass of the reaction mixture ($N_I = M/\overline{m}_I$), which remains constant in the course of the process. This is not the case in a SBR, because of the effect of the inlet/outlet streams. For the same reason, the choice of the initial mass or volume as a reference will not simplify the formulation either.

Energy conservation equation:

The energy conservation equation here used can be obtained from the difference between the total energy balance and the mechanical energy balance, whose details can be seen in, e.g., [6]. After rearrangement of the resulting *thermal energy balance*, considering uniform properties in the reacting mixture and neglecting the viscous dissipation effects and thermal expansion of the liquid mixture:

$$M\hat{c}_p \frac{dT}{dt} = \sum_{\alpha} W_{\alpha}[\hat{c}_{p,\alpha}(T_{\alpha} - T) - \Delta \hat{H}^d_{\alpha}] + V \sum_k (-\Delta H_k) r_k - Q^*; \qquad T(0) = T_I \tag{2.92}$$

In Eq. (2.92), T is the instantaneous temperature of the liquid mixture inside the tank, $\hat{c}_p$ and ΔH_k are the values of the specific heat capacity and of the reaction enthalpy of the kth canonical reaction, $\hat{c}_{p,\alpha}$ is the specific heat capacity of the stream α averaged between temperatures T_α and T, Q^* is the rate of heat transferred from the tank to a heat-exchange device, and $\Delta \hat{H}^d_{\alpha}$ is the heat of mixing from the dissolution of the stream α in the tank solution, evaluated at temperature T, which can be written in terms of the specific heat of mixing of each species A_j, $\Delta \hat{h}^d_{j\alpha}$:

$$\Delta \hat{H}^d_{\alpha} = \sum_j \omega_{j\alpha} \Delta \hat{h}^d_{j\alpha} \tag{2.93}$$

2.8.1 Stoichiometric Relationships for the Number of Moles N_j

Having chosen a set of K key species, it is recalled that the production rates of each component species is given by $r_j = \sum_k \sigma_{kj} r_k$ $(j > K)$. After replacing in this relationship the values of r_j and r_k from the mass conservation equation for each species (Eq. 2.91):

$$\frac{dN_j}{dt} + \frac{N_j}{\theta} - \sum_{\alpha} F_{j\alpha} = \sum_k \sigma_{kj} \left(\frac{dN_k}{dt} + \frac{N_k}{\theta} - \sum_{\alpha} F_{k\alpha} \right); \qquad j > K \tag{2.94}$$

Multiplying Eq. (2.94) by the following integrating factor,

$$E(t) = \exp\left(\int_0^t \frac{dt}{\theta} \right) = \exp\left(\int_0^t \frac{W}{M} dt \right) \tag{2.95}$$

and rearranging:

$$\frac{d[E(t)N_j]}{dt} - E(t) \sum_{\alpha} F_{j\alpha} = \sum_k \sigma_{kj} \left[\frac{d[E(t)N_k]}{dt} - E(t) \sum_{\alpha} F_{k\alpha} \right]; \qquad j > K$$

Upon integration between the initial time ($t = 0$) and a generic value t, it results:

$$\Delta N_j = \sum_k \sigma_{kj} \Delta N_k; \qquad j > K \tag{2.96}$$

where for any key or component species:

$$\Delta N_j = N_j - N_j^*; \qquad N_j^* = E^{-1}(t)\left[N_{jI} + \int_0^t E(t)\left(\sum_\alpha F_{j\alpha}\right)dt\right] \tag{2.97}$$

The evaluation of $E(t)$ requires expressing $\theta = M/W$ (see Eq. 2.95). This can be accomplished by integration of Eq. (2.87) for M:

$$\theta = \frac{1}{W}\left[M_I + \int_0^t \left(\sum_\alpha W_\alpha - W\right)dt\right] \tag{2.98}$$

Equations (2.96) and (2.97) gives the stoichiometric relationships that can replace the mass conservation Eq. (2.91) for the component species, while keeping Eq. (2.91) for the K key species A_k:

$$\frac{dN_k}{dt} = \sum_\alpha F_{k\alpha} - N_k/\theta + V r_k; \qquad N_k(0) = N_{kI} \tag{2.99}$$

A significant simplification arises when no homogeneous extraction exists ($W = 0$). Thus, it is obtained from Eq. (2.95) $\boldsymbol{E(t)} = 1$, and hence the expression of N_j^* (Eq. 2.97) becomes:

$$(W = 0) \qquad N_j^* = N_{jI} + \int_0^t \left(\sum_\alpha F_{j\alpha}\right)dt; \qquad j = 1, \ldots, S \tag{2.100}$$

In such a case, N_j^* coincide with the net number of moles of A_j fed into the reactor since the operation start-up.

The evaluation of the N_j^*, either from Eq. (2.97) or (2.100), requires specifying the flow rates and composition of the inlet/outlet streams to calculate the term $\sum_\alpha F_{j\alpha} = \sum_\alpha(\omega_{j\alpha} W_\alpha/m_j)$. In addition, if the homogeneous stream is involved (Eq. 2.97 must be used for N_j^*), the mass flow rate W should be specified too.

If the above-mentioned flow rates are established beforehand, eventually as a function of time t, the N_j^* can be evaluated independently of the evolution of the system and Eq. (2.97) will allow to explicitly evaluate the instantaneous values N_j for given values N_k of the key species. Subsequently, these relationships can be used in the resolution of the key-species balances (2.99), similarly as for, e.g., the BR. Nonetheless, it is worth noting that relationships (2.96) involve the effect of the inlet/outlet streams (by the way of the N_j^*), a distinct aspect from the relationships arisen in previous sections.

On the contrary, if any of the flow rates W or $F_{j\alpha}$ are subject to the composition and/or temperature changes of the reaction mixture (i.e., to N_j and/or T changes), Eqs. (2.96), (2.99) and eventually (2.92) for T must, in general, be solved simultaneously. Then, there will be no evident advantage from using Eq. (2.96) for the component species instead of their balances, Eq. (2.91).

Considering that the inlet/outlet streams other than the homogeneous extraction in a SBR will be frequently employed to automatically control the evolution of the system, it is probable that, in many practical situations, the evaluation of the N_j^* in advance cannot be accomplished.

The streams that are not pre-set will be referred to as *control-streams* for briefness. As an example, the dosed addition of a reactant can aim at maintaining its concentration in an appropriate level, e.g., for selectivity reasons or to moderate the heat production rate. In such a case, the inlet flow rate will be conditioned by the consumption rate of the reactant and therefore will depend on the evolution of the system.

Another example arises when a given product is to be continuously removed from the reactor (e.g., to increase the conversion in equilibrium-limited reactions) by means of a selective membrane or by vapour-phase extraction when the product presents a higher volatility than the rest of species. Alternatively, for the same purpose, a homogeneous stream can be extracted from the reactor, processed in an external equipment to separate the desired product and the remainder recycled to the reactor. Whichever the chosen separation process is, only the net extraction can be accounted for in the mass conservation equations.

For either alternative, to efficiently satisfy the objective of the control action, the extraction rate will be subjected to the chemical production rate of the extracted species, and hence to the instantaneous state of the reaction mixture.

To explicitly visualise this feature, consider a vapour-phase extraction with the control target of keeping the mole fraction of the volatile product P at a certain level y_P. The balance of P (Eq. 2.91), assuming that $W = 0$ and that P is not involved in any other stream, is $dN_P/dt = V\mathscr{r}_P + F_{Pv}$ where F_{Pv} is the molar flow rate of P in the extraction stream (negative, as mentioned at the beginning of Sect. 2.8). As $N_P = N y_P$ and with constant y_P:

$$F_{Pv} = -V\mathscr{r}_P + y_P(dN/dt) \tag{2.101a}$$

In the vapour stream, $F_{Pv} = y_{Pv} F_v$, where y_{Pv} is the mole fraction of P in the vapour stream and F_v is the vapour molar flow rate. Assuming vapour–liquid equilibrium $y_{Pv} = y_P K_P$, where K_P is the vapour–liquid equilibrium constant:

$$F_{Pv} = y_P K_P F_v \tag{2.101b}$$

From Eqs. (2.101a, b), the vapour molar flow rate F_v is expressed as:

$$F_v = [-V\mathscr{r}_P + y_P(dN/dt)]/(y_P K_P) \tag{2.101c}$$

Equation (2.101c) clearly shows the dependency of F_v on the instantaneous state of the reacting mixture through the variables V, $\mathscr{r}_P$, dN/dt and K_P.

It is usual that only a low number of control-streams and their associated species are implemented to operate a SBR. If the remainder species are not involved in such control-streams and additionally $W = 0$ (a common case), their reference values will be given by (Eq. 2.100) as $N_j^* = N_{jI}$. Provided that N_k key species can be identified within this group, for the other species in the group (secondary species) the stoichiometric Eq. (2.96) simply read: $N_j = N_{jI} + \sum_k \sigma_{kj}(N_k - N_{kI})$. Therefore, the use of these expressions will practically reduce the number of equations to solve simultaneously (those of the key species and of the control-involved species). Actually, the procedure can be extended to the case when the group of "non-controlled" species are involved in pre-set inlet streams (see Eq. 2.100).

Example 1: Analysis of a SBR for processing an exothermic equilibrium-limited reaction

The compound C *is obtained from the liquid-phase reaction between* A *and* B, *according to:*

$$\mathcal{R}_1: \quad \mathrm{A + B \Leftrightarrow C + D}; \qquad r_1 = k_1(C_\mathrm{A} C_\mathrm{B} - C_\mathrm{C} C_\mathrm{D}/K_{C1})$$

$$\mathcal{R}_2: \quad \mathrm{B + 2C \Rightarrow E + S}; \qquad r_2 = k_2 C_\mathrm{B} C_\mathrm{C}^2$$

Given that both reactions are highly exothermic, and the desired reaction is strongly limited by the chemical equilibrium, the following protocol is proposed to operate <u>*isothermally*</u> *in a SBR:*

- *The reactor is initially charged with a volume V_I of a solution containing a solvent and only the reactant* A *at a molar concentration $C_{\mathrm{A}I}$.*
- *To moderate the heat production, while simultaneously restraining the undesired side reaction, the reactant* B *is dosed from a liquid solution (stream α) at a given constant volumetric flow rate q_α and molar concentration $C_{\mathrm{B}\alpha}$, during a pre-set time t_α.*
- *To displace the chemical equilibrium towards the products,* C *is continuously extracted using a highly selective membrane that allows keeping its molar concentration in the liquid phase at a known and almost negligible level, $C_{\mathrm{C},max}$, during most of the operation.*

Additional assumptions:

- *The dependence of kinetic and chemical equilibrium parameters on T is known, as well as the heat of reactions.*
- *Physical properties such as ρ, ρ_α and $\hat{c}_{p,\alpha}$ are known and can be regarded as constant.*
- *The amount of the remaining species passing through the membrane can be neglected.*

It is required to express the final form of the equation set that should be solved to analyse the operation during the time t_α, including the expression of the instantaneous heat transfer rate Q^* needed to keep the operation at a chosen constant temperature T.

Solution:

Species A and E are chosen as key component (then $\sigma_{\mathrm{AB}} = 1$, $\sigma_{\mathrm{AC}} = -1$, $\sigma_{\mathrm{AD}} = -1$, $\sigma_{\mathrm{AS}} = 0$, and $\sigma_{\mathrm{EB}} = -1$, $\sigma_{\mathrm{EC}} = -2$, $\sigma_{\mathrm{ED}} = 0$, $\sigma_{\mathrm{ES}} = 1$). Equation (2.91) for A and E leads to:

$$dN_A/dt = -Vr_1; \qquad N_A(0) = N_{AI} \tag{E2.1}$$

$$dN_E/dt = Vr_2; \qquad N_E(0) = N_{EI} = 0 \tag{E2.2}$$

Given that the species C is involved in a control action that depends on the reaction progress, it is omitted from the stoichiometric relationships. On the other hand, as there is no homogeneous extraction, Eqs. (2.96) and (2.100) give for the component species B, D and S (with $N_{BI} = N_{DI} = N_{SI} = N_{EI} = 0$):

$$N_B = (N_A - N_{AI}) - N_E + q_\alpha C_{B\alpha} t \tag{E2.3}$$

$$N_D = -(N_A - N_{AI}) \tag{E2.4}$$

$$N_S = N_E \tag{E2.5}$$

Considering that the molar concentration of species C should be kept at a very low level $C_{C,max}$, $dN_C/dt \approx 0$ can be considered. Then, from the conservation Eq. (2.91) for species C:

$$dN_C/dt = F_{C,m} + V(r_1 - 2r_2) \approx 0 \quad \Rightarrow \quad F_{C,m} = -V(r_1 - 2r_2) \tag{E2.6}$$

where $F_{C,m}$ is the instantaneous molar flow rate of C that should be extracted through the membrane.

The mass conservation Eq. (2.88), with $\sum_\alpha W_\alpha = \rho_\alpha q_\alpha + m_C F_{C,m}$, considering the density ρ as constant and taking into account Eq. (E2.6) yields

$$\rho dV/dt = \rho_\alpha q_\alpha - m_C V(r_1 - 2r_2); \qquad V(0) = V_I \tag{E2.7}$$

As a further simplification, Eqs. (E2.1) and (E2.2) can be used to replace the terms Vr_1 and Vr_2 in Eq. (E2.7), which after integration results in:

$$\rho(V - V_I) = \rho_\alpha q_\alpha t - m_C[(N_{AI} - N_A) - 2N_E] \tag{E2.8}$$

In addition, Vr_1 and Vr_2 can be explicitly expressed with the help of Eq. (E2.3)–(E2.5) and $C_C = C_{C,max}$:

$$Vr_1 = k_1\left\{\frac{N_A[(N_A - N_{AI}) - (N_E - N_{EI}) + q_\alpha C_{B\alpha} t]}{V} - \frac{C_{C,max}(N_{AI} - N_A)}{K_{C1}}\right\} \tag{E2.9}$$

$$Vr_2 = k_2[(N_A - N_{AI}) - (N_E - N_{EI}) + q_\alpha C_{B\alpha} t]C_{C,max}^2 \tag{E2.10}$$

Finally, to express Q^* during the operation time, it is assumed that the extraction of species C does not involve any significant thermal effect. Then, from the energy conservation Eq. (2.92), with $dT/dt = 0$:

$$Q^* = q_\alpha \hat{c}_{p,\alpha}(T_\alpha - T) + (-\Delta H_A)V(-r_1) + (-\Delta H_E)Vr_2 \tag{E2.11}$$

The set of equations to be solved for the state variables are the two differential equations (E2.1), (E2.2) and the algebraic equations (E2.3)–(E2.5) and (E2.8). Their resolution will allow to evaluate instantaneous values of the extraction flow rate $F_{C,m}$ (Eq. E2.6) and of the heat-exchange rate Q^* (Eq. E2.11).

In actual practice, the control-streams can contain minor amounts of the remaining species that may be required to be considered in the evaluation of the reactor operation. It is discussed in the Appendix 3 that the stoichiometric relationships could be efficiently used in such cases, even in operations with a homogeneous extraction ($W \neq 0$). However, it is considered that the treatment presented in Appendix 3 is beyond the scope of an introductory CRE course and is therefore suggested as further reading.

2.8.2 Application to CSTRs Start-Up

The start-up of a CSTR, under the same conditions previously considered for the steady-state analysis of Sect. 2.6, is addressed here. Whichever the start-up routine, the expressions discussed in Sect. 2.8 can be applied to simulate the time evolution of the reactor. Several feed streams may be considered, but only one extraction stream from the tank content. In a general case, the reactor can be in a process operation train and then the feeding conditions should probably depend on the conditions of other upstream process units. In such a case, even though the initial conditions could be clearly specified, the feeding conditions (flow rates, temperature, composition) may not be defined in advance. Thus, it can be concluded that the stoichiometric relationships (2.96) will no longer be of practical use.

Nevertheless, the situation when the feeding conditions are fixed according to the specified steady state (SS) is of interest for emphasising the importance of reactor start-up, especially for didactic purposes. To this end, the specific form of the conservation equations after imposing certain simplifications, typically considered in the treatment of stirred tank reactors, together with the application of the stoichiometric relationships is described in this section. In this way, it will be possible to analyse the impact of initial conditions on the system evolution until the SS is reached.

Along with the consideration of constant feeding conditions, it will be assumed that the tank is equipped with a level controller to keep the liquid volume V constant during the start-up, i.e., $V = V_I$ for all time t.

Then, defining as W_0 the total inlet mass flow rate, $W_0 = \sum_\alpha W_\alpha$, the total mass conservation equation, Eq. (2.88), results in:

$$dM/dt = V d\rho/dt = W_0 - q\rho; \qquad M(0) = M_I \tag{2.102}$$

Simplification 1: It is assumed that the density of the liquid mixture remains unchanged, $\rho = \rho_I$. Then, from Eq. (2.102), $W_0 = q\rho$, which allows concluding that q, M and $\theta = V/q = M/W_0$ will also remain constant.

Regarding the mass conservation equation for each species, the invariance of V and ρ suggests rewriting it in terms of molar concentration $C_j = N_j/V$. Particularly for a key species (Eq. 2.99):

$$dC_k/dt = (C_{k0} - C_k)/\theta + r_k; \qquad C_k(0) = C_{kI} \tag{2.103}$$

where in general, as defined in Sect. 2.6, C_{j0} is the hypothetical inlet molar concentration of A_j defined as:

$$C_{j0} = \left(\sum\nolimits_\alpha F_{j\alpha}\right)/q \tag{2.104}$$

For the component species, the stoichiometric relationships (Eqs. 2.96 and 2.97) are also expressed in terms of C_j. Furthermore, as $\theta = V/q$ is constant, the integrating factor given in Eq. (2.95) results in $E(t) = e^{t/\theta}$. Then, from Eqs. (2.96) and (2.97):

$$C_j = e^{-t/\theta}\mathbb{C}_{jI} + \left(1 - e^{-t/\theta}\right)\mathbb{C}_{j0} + \sum_k \sigma_{kj} C_k; \qquad j > K \tag{2.105}$$

where

$$\mathbb{C}_{jI} = C_{jI} - \sum_k \sigma_{kj} C_{kI} \tag{2.106a}$$

$$\mathbb{C}_{j0} = C_{j0} - \sum_k \sigma_{kj} C_{k0} \tag{2.106b}$$

It can be observed from Eq. (2.105) that the quantity $\mathbb{C}_j = C_j - \sum_k \sigma_{kj} C_k = e^{-t/\theta}\mathbb{C}_{jI} + \left(1 - e^{-t/\theta}\right)\mathbb{C}_{j0}$ $(j > K)$, based on molar concentrations, is independent of any effect due to chemical reactions. Then, as defined in Sect. 2.2, the $\mathbb{C}_j$ behave as *reaction invariants*, depending explicitly on time according to weighted averages between the initial component variables $\mathbb{C}_{jI}$ and the inlet component variables $\mathbb{C}_{j0}$. Moreover, at high enough times, Eq. (2.105) lead to the SS stoichiometric relationships (Eq. 2.66b):

$$(\text{SS}): \quad C_{j0} = \mathbb{C}_{j0} + \sum_k \sigma_{kj} C_k = C_{j0} + \sum_k \sigma_{kj}(C_k - C_{k0}); \qquad j > K \tag{2.107}$$

Additionally, when for some A_j the initial value $\mathbb{C}_{jI}$ is equal to $\mathbb{C}_{j0}$, the variable $\mathbb{C}_j$ becomes independent of time and Eq. (2.107) can be used for the component species over the whole time-evolution towards the SS.

To consider the thermal energy balance, Eq. (2.92), the following definitions concerning the inlet stream are first introduced:

$$\hat{c}_{p0} = \left(\sum\nolimits_{\alpha} W_{\alpha}\hat{c}_{p,\alpha}\right)/W_0 \tag{2.108a}$$

$$T_0 = \left(\sum\nolimits_{\alpha} W_{\alpha}\hat{c}_{p,\alpha}T_{\alpha}\right)/(W_0\hat{c}_{p0}) \tag{2.108b}$$

$$\Delta T^d = \left[\sum\nolimits_{\alpha} W_{\alpha}(-\Delta\hat{H}_{\alpha}^d)\right]/(W_0\hat{c}_{p0}) \tag{2.108c}$$

Then, Eq. (2.92) can be written as:

$$M\hat{c}_p\frac{dT}{dt} = W_0\hat{c}_{p0}\left[(T_0 - T) + \Delta T^d\right] + V\sum_k(-\Delta H_k)r_k - Q^*; \qquad T(0) = T_I \tag{2.109}$$

For an adiabatic operation is also possible to relate temperature and composition during the start-up. Taking $Q^* = 0$ in Eq. (2.109) and replacing the r_k from Eq. (2.103) for the key species:

$$M\hat{c}_p\frac{dT}{dt} = W_0\hat{c}_{p0}\left[(T_0 - T) + \Delta T^d\right] + V\sum_k(-\Delta H_k)\left[\frac{dC_k}{dt} - \frac{(C_{k0} - C_k)}{\theta}\right] \tag{2.110}$$

To obtain a simple relationship, additional assumptions regarding the thermal parameters are further needed. These are identified as *simplifications 2* and *3* below.

Simplification 2: It is assumed that $\hat{c}_{p0} = \hat{c}_p$.

Then, upon dividing Eq. (2.110) by $M\hat{c}_p$:

$$\frac{dT}{dt} = \frac{1}{\theta}\left[\left(T_0 + \Delta T^d\right) - T\right] + \sum_k J_k\left[\frac{dC_k}{dt} - \frac{(C_{k0} - C_k)}{\theta}\right] \tag{2.111}$$

where $J_k = (-\Delta H_k)/(\rho\hat{c}_p)$.

Simplification 3: It is further assumed that J_k and ΔT^d are constant.

Considering this simplification, and after multiplying by the integrating factor $E(t) = e^{t/\theta}$, Eq. (2.111) can now be integrated, resulting in:

$$T = e^{-t/\theta}\mathbb{T}_I + \left(1 - e^{-t/\theta}\right)\mathbb{T}_0 + \sum_k J_kC_k \tag{2.112}$$

where,

$$\mathbb{T}_I = T_I - \sum_k J_kC_{kI} \tag{2.113a}$$

$$\mathbb{T}_0 = (T_0 + \Delta T^d) - \sum_k J_kC_{k0} \tag{2.113b}$$

From Eq. (2.112), it can be concluded that the quantity defined as $\mathbb{T} = T - \sum_k J_kC_k$ does not depend on any effect of the chemical reactions, so that it behaves as a *thermal reaction invariant*. The time dependency of $\mathbb{T}$ results in a weighted average between the initial value $\mathbb{T}_I$ and the inlet value $\mathbb{T}_0$. Then, for an adiabatic operation, the differential Eq. (2.103) must be integrated together with the algebraic Eqs. (2.105) and (2.112).

When the SS is reached at high enough times, Eqs. (2.112) and (2.113b) lead to the following relationship (cfr. Eq. 2.72):

$$\text{(SS):} \quad T = T_0 + \Delta T^d + \sum_k J_k(C_k - C_{k0}) \tag{2.114}$$

2.9 Conclusions

A series of elementary single-phase reacting systems as listed in the introduction, and typically covered in introductory CRE courses, were analysed throughout this chapter, with the emphasis on the use of the stoichiometric relationships between the net production rates of species, $r_j = \sum_k \sigma_{kj} r_k \ (j > K)$. It was shown that those relationships can be conveniently used to establish equivalent relationships between the amount of the component species and key species present in the reacting system. This allows reducing the number of conservation equations that need to be solved and simultaneously recognising the number of independent variables that should be fixed to define the state of the system (i.e., the number of *degrees of freedom*).

For most reaction systems in this chapter, those relationships could be described in terms of extensive variables, such as molar flow rates or the number of moles of the species. However, the convenience of employing intensive variables was also discussed in Sect. 2.2, for which appropriate measures of concentration must be chosen. Mass fractions, ω_j, or moles per unit total mass, ψ_j, were identified as possible alternatives. Nonetheless, it was discussed that both alternatives show some undesirable aspects from a didactic point of view. Instead, the use of *relative variables*, defined as the ratio between the species extensive variables (e.g., molar flow rates, F_j) and some other extensive reference variable, was found to be a convenient option.

Particularly, the relative variable $\mathcal{Y}_j$, defined as the moles of A_j per total number of moles of a given reference (the total inlet molar flow rate for continuous-flow reactors, or the initial total number of moles in the liquid mixture for a batch reactor), was considered as the most suitable one. The conservation equations and stoichiometric relationships were conveniently formulated in terms of these relative variables for all systems analysed, except for the *semi-batch reactor* (Sect. 2.8). This reactor presents operation features not found in the other studied cases. As such, it imposes limitations for the practical use of the stoichiometric relationships between the amounts of the species. Although such limitations can be removed under certain operating conditions, the stoichiometric relationships involve a time dependence, which is a distinct feature from all other reaction systems in this chapter.

Finally, the concept of *reaction-invariant* was introduced in Sect. 2.2. These variables are associated with an extensive (e.g., F_j) or intensive (e.g., ω_j, ψ_j) measure of the amount present of the component species, for which the stoichiometric relationships can be established. The reaction invariants remain independent of the reaction progress and allow writing such relationships in an alternative way. Nevertheless, it was mentioned that this concept is not necessary for an introductory CRE course. Indeed, it was not used throughout the chapter, except for the analysis of the start-up of a CSTR and the treatment of semi-batch reactors presented in the Appendix 3 as further reading. Yet, even in these cases, the formal definition of a reaction invariant could be avoided. It is finally mentioned that the concept and usefulness of reaction invariants will be evident in Chap. 3, where more complex reaction systems and models will be analysed.

Appendix 1: Relationships Between Variables Ω_j and $\mathbb{Y}_j$

The relationships between mass and mole fractions are first recalled:

$$\omega_j = (y_j m_j)/\overline{m} \tag{2.115}$$

where the mean molar mass, $\overline{m}$, is:

$$\overline{m} = \sum_j y_j m_j$$

The definitions of $\mathbb{Y}_j$ (Eq. 2.13a) and Ω_j (Eq. 2.23a) are here repeated:

$$\Omega_j = \omega_j - \sum_k \hat{\sigma}_{kj}\omega_k; \qquad \mathbb{Y}_j = \frac{y_j - \sum_k \sigma_{kj} y_k}{1 - \sum_k \Delta\sigma_{kj} y_k}$$

Replacing Eq. (2.115) in the definition of Ω_j and expressing Ω_j/m_j by taking into account that $\hat{\sigma}_{kj} = \frac{m_j}{m_k}\sigma_{kj}$:

$$\Omega_j/m_j = \left(y_j - \sum_k \sigma_{kj} y_k\right)/\overline{m} \tag{2.116}$$

Summing up Eq. (2.116) for all component species ($j > K$), and bearing in mind that $\sum_{j>K}\sigma_{kj} = \Delta\sigma_k - 1$:

$$\sum_{j>K} \Omega_j/m_j = \left(1 - \sum_k \Delta\sigma_k y_k\right)/\overline{m} \tag{2.117}$$

From the term-by-term division between Eqs. (2.116) and (2.117):

$$\mathbb{Y}_j = \frac{\Omega_j / m_j}{\sum_{\beta > K} \Omega_\beta / m_\beta} \tag{2.118}$$

Finally, recalling that $\sum_{j>K} \Omega_j = 1$, it can also be obtained from Eq. (2.118) the reciprocal relationship:

$$\Omega_j = \frac{m_j \mathbb{Y}_j}{\sum_{\beta > K} m_\beta \mathbb{Y}_\beta} \tag{2.119}$$

Noted also that, from $\Omega_j = m_j \Psi_j$ (Eq. 2.26a) and Eq. (2.118), $\mathbb{Y}_j$ and Ψ_j are related by:

$$\mathbb{Y}_j = \frac{\Psi_j}{\sum_{\beta > K} \Psi_\beta} \tag{2.120}$$

Appendix 2: Analysis of the $K_{C,k}$ Constants in Liquid Phase

We will consider here the particular case of liquid-phase reactions in equilibrium under the assumption of constant density. Therefore, the use of $K_{C,k}$ in terms of $\Delta G_k^0(P, T)$ (Table 2.1) will be assumed.

It was argued in Sect. 2.4 that, even if the solution is assumed to behave as an ideal solution ($K_{\gamma,k} = 1$),$K_{C,k}$ will still depend on the unknowns C_k if the reactions do not conserve the number of moles ($\Delta\sigma_k \neq 0$), due to the term $\upsilon^{\Delta\sigma_k} = C^{-\Delta\sigma_k}$ (see Eq. 2.39b). This effect is frequently ignored when the ideal solution is invoked, even though it would be significant if the reactants are concentrated. Nevertheless, as already discussed in Sect. 2.4, the fact that $\Delta\sigma_k \neq 0$ almost invariably implies that the molecular sizes of the species are different. In such a case, the solution should not be expected to behave as ideal, even if the molecules do not differ significantly in their relative attractive forces. Such an effect is explicitly recognised in a number of activity coefficients models, like UNIFAC, UNIQUAC, ASOG, MOSCED, among others, any of which expresses:

$$\gamma_j = \gamma_{comb,j} \gamma_{res,j} \tag{2.121}$$

where $\gamma_{comb,j}$ quantifies the differences in size and shape of the molecules (the so-called *combinatorial contribution*), requiring only pure-component data, while $\gamma_{res,j}$ primarily accounts for the differences in the attractive forces between the distinct molecules (the so-called *residual contribution*) [5].

For the equilibrium constant $K_{C,k}$, as written in Table 2.1, $K_{\gamma,k}$ can be expressed from Eq. (2.121) as:

$$K_{\gamma,k} = K_{comb,k} K_{res,k} \tag{2.122}$$

where $K_{comb,k} = \gamma_{comb,k}\prod_{j>K}\gamma_{comb,j}^{\sigma_{kj}}$, and $K_{res,k} = \gamma_{res,k}\prod_{j>K}\gamma_{res,j}^{\sigma_{kj}}$

The simplest expression to reflect the main effect of the difference in the molecular sizes is the Flory–Huggins type, from which $\gamma_{comb,j}$ may be written in the following form (see, e.g., [7]):

$$\gamma_{comb,j} = (\upsilon_j^0/\upsilon^0)\exp[1-(\upsilon_j^0/\upsilon^0)] \tag{2.123}$$

$$\upsilon^0 = \sum_j y_j\upsilon_j^0 \tag{2.124}$$

where υ_j^0 is the molar volume of the pure A_j at the temperature T and pressure P of the system, quantifying the size of the A_j species. From a strictly theoretical point of view, the size of each molecule should be evaluated by a magnitude independent of T and P (e.g., using the standard values, $T = 25°C$ and $P = 1atm$). However, considering the weak variation of the molar volume with T and P and that the effect is introduced in relative terms, υ_j^0/υ^0, the use of υ_j^0 can be regarded as adequate.

From Eqs. (2.122) and (2.123), $K_{C,k}$ can then be written as:

$$K_{C,k} = \frac{K_k^0}{K_{res,k}e^{\Delta\sigma_k}K_{\upsilon,k}^0\exp(-\Delta\upsilon_k^0/\upsilon^0)}(\upsilon^0/\upsilon)^{\Delta\sigma_k} \tag{2.125}$$

where

$$K_{\upsilon,k}^0 = \upsilon_k^0\prod_{j>K}(\upsilon_j^0)^{\sigma_{kj}} \tag{2.126}$$

$$\Delta\upsilon_k^0 = \upsilon_k^0 + \sum_{j>K}\sigma_{kj}\upsilon_j^0 \tag{2.127}$$

The impacts of the assumption of constant ρ during the course of the reactions on $K_{C,k}$ can now be addressed based on Eq. (2.125).

The change in the specific volume of the solution $\hat{\upsilon} = 1/\rho$, at constant temperature, can be conveniently written as $d\hat{\upsilon} = \sum_j \upsilon_j d\psi_j$, where υ_j is the partial molar volume of each species A_j in the mixture and ψ_j is the variable defined in Sect. 2.2 (moles of A_j per unit mass of solution). The effect of the pressure change is ignored in such expression, considering that liquids behave as incompressible unless the change in P is exceptionally high. Using Eq. (2.16) for the ψ_j values of the component species ($j > K$) and rearranging:

$$d\hat{\upsilon} = \sum_k \Delta\upsilon_k d\psi_k \tag{2.128}$$

where $\Delta\upsilon_k$ is the change in the molar volume promoted by the kth canonical reaction:

$$\Delta \upsilon_k = \upsilon_k + \sum_{j>K} \sigma_{kj} \upsilon_j \tag{2.129}$$

Under the constant ρ assumption, it must be $d\hat{\upsilon} = 0$. According to Eq. (2.128), this implies $\Delta \upsilon_k = 0$, since the variation of the $d\psi_k$ is independent of each other. Thus, from the definition in Eq. (2.129):

$$\upsilon_k + \sum_{j>K} \sigma_{kj} \upsilon_j = 0 \tag{2.130}$$

Due to the fact that Eq. (2.130) must be verified in the whole reaction course, it can be reasoned that for Eq. (2.130) to be true, the partial molar volume υ_j of each species must be kept constant. It should be noted that the latter is a two-way condition: if the partial molar volumes depend on the composition, Eq. (2.130) cannot be expected to hold as reactions progress and, therefore, the density cannot be strictly constant.

Hence, with the constant ρ assumption, $\upsilon_j = \upsilon_{j0}$ can be taken for consistency and Eq. (2.130) should then be satisfied by:

$$\upsilon_{k0} + \sum_{j>K} \sigma_{kj} \upsilon_{j0} = 0 \tag{2.131}$$

Furthermore, from $\upsilon = \sum_j y_j \upsilon_j$:

$$\upsilon = \sum_j y_j \upsilon_{j0} \tag{2.132}$$

Considering Eq. (2.124) for υ^0 and Eq. (2.132) for υ, it can be expected for the quotient (υ^0/υ) in Eq. (2.125) to be close to 1, since any significant departure would arise from unlikely significant differences between the molar volumes of the pure species υ_j^0 and the partial molar volumes in the mixture, υ_{j0}. In addition, if Eq. (2.132) could be extended to the limits of pure substances, $\upsilon = \upsilon^0$. Therefore, it can be reasonably assumed that $\upsilon^0/\upsilon \approx 1$ in Eq. (2.125). In a similar fashion, by virtue of Eq. (2.131) and considering that the υ_{j0} are closed enough to the υ_j^0, it can be concluded that $\Delta\upsilon_k^0 \approx 0$.

Then, from Eq. (2.125), the following approximate expression arises for $K_{C,k}$:

$$K_{C,k} = \frac{K_k^0}{K_{res,k} e^{\Delta\sigma_k} K_{\upsilon,k}^0} \tag{2.133}$$

Equation (2.133) shows that, after due consideration of the activity coefficients and consistency with the constant-density assumption, the direct effect of υ in the expression of $K_{C,k}$ in Table 2.1, at least approximately, cancels out.

It is worth noting that $K_{C,k}$ will still depend on the equilibrium composition (i.e., on the unknowns C_k) through the residual term, $K_{res,k}$, when the attractive forces between different molecules exhibit significant differences.

Finally, a last observation must be made. The fact that Eq. (2.130) applies to every reaction suggests that a relationship between the partial molar volume of the reacting species, υ_j, should exist. The simplest relationship is $\upsilon_j = m_j/\rho$, i.e., when each molar volume is proportional to the molar mass. In such a case, Eq. (2.130) will be automatically satisfied for every reaction, since $m_k + \sum_{j>K} \sigma_{kj} m_j = 0$.

Appendix 3: Use of Stoichiometric Relationships When SBRs Are Operated with Control-Streams

Open reactors with perfect mixing under unsteady state, or simply *semi-batch reactors*, were considered in Sect. 2.8. The frequently found situation when control actions are performed by the addition or removal of certain species in *control-streams* was addressed there. It was discussed that in such a case the practical use of the general stoichiometric relationships involving Eqs. (2.95)–(2.97) are highly restricted because of the dependence of the reference values N_j^* on the reaction rates. The case when the control-streams of certain species strictly exclude the remaining of the species and there is no homogeneous extraction ($W = 0$) was pointed out as an exception allowing the effective use of the stoichiometric relationships for the remaining species.

The applicability of the stoichiometric relationships in a more general case in which the control-streams may involve a small amount of the remaining species will be discussed here. In addition, it will also be considered the presence of a homogeneous extraction with a prescribed mass flow rate, $W(t)$.

We start by defining new variables associated with N_j and $F_{j\alpha}$ of the component species:

$$\mathbb{N}_j = N_j - \sum_k \sigma_{kj} N_k; \qquad j > K \tag{2.134}$$

$$\mathbb{F}_{j\alpha} = F_{j\alpha} - \sum_k \sigma_{kj} F_{k\alpha}; \qquad j > K \tag{2.135}$$

The variables $\mathbb{N}_j$ and $\mathbb{F}_{j\alpha}$ are component variables, as defined in Sect. 2.2.1. In particular, $\mathbb{N}_j$ can be interpreted as the number of moles of A_j potentially attainable if the N_k moles of the key species were totally consumed by the reactions involving A_j.

From these definitions, Eq. (2.94) in Sect. 2.8.1 can be rewritten as:

$$\frac{d\mathbb{N}_j}{dt} + \frac{\mathbb{N}_j}{\theta} = \sum_\alpha \mathbb{F}_{j\alpha}; \qquad \mathbb{N}_j(0) = \mathbb{N}_{jI}; \qquad j > K \tag{2.136}$$

where the initial condition $\mathbb{N}_{jI} = N_{jI} - \sum_k \sigma_{kj} N_{kI}$ is included in Eq. (2.136).

Equation (2.136) can substitute the mass conservation equations of the component species (Eq. 2.91). Thus, the set of species balances will be defined by Eq. (2.99) for the key-species values N_k and Equation (2.136) for the component variables $\mathbb{N}_j$. It is noted that knowing $\mathbb{N}_j$ and N_k, the number of moles of the component species is evaluated in an elementary way from Eq. (2.134):

$$N_j = \mathbb{N}_j + \sum_k \sigma_{kj} N_k; \qquad j > K \tag{2.137}$$

The potential advantage of using Eq. (2.136) will be analysed in the frequent case that only a small number E of species are involved in control actions. Such species will be identified as A_e, with index e ranging from $K + 1$ to $K + E$. It will be assumed that within the remainder species a number K of them can be identified as key species and $(S - K - E)$ as component species A_j $(j > K + E)$, and that all of them could be present only marginally in the control-streams.

Non-homogeneous streams (identified by the index α in Sect. 2.8) are discriminated into two groups:

- A number B of inlet streams, identified by the index b $(1 \leq b \leq B)$, with a prescribed molar flow rates $F_b(t)$ and constant mole fractions y_{jb}. They may or may not have any of the A_e species, with no further distinction needed. In practice, the number B of these streams will be typically low, say 1 or 2.
- A number C of control-streams, identified by the index c $(B + 1 \leq c \leq B + C)$, which are supposed to contain any of the E species involved in control actions ($C = E$ can be expected to hold as a common situation) and only marginal amounts of the remaining species. The molar flow rates F_{jc} of these control-streams as such will depend on the instantaneous values of the state variables (T, P, and composition in the liquid mixture).

As already mentioned, if a homogeneous extraction is considered, a prescribed mass flow rate $W(t)$ will be assumed.

By taking into account the discrimination made between the inlet/outlet streams, Eq. (2.136) for the component species that are not the subject of control actions are rewritten as:

$$\frac{d\mathbb{N}_j}{dt} + \frac{\mathbb{N}_j}{\theta} = \sum_b \mathbb{y}_{jb} F_b + \sum_c \mathbb{F}_{jc}; \qquad \mathbb{N}_j(0) = \mathbb{N}_{jI}; \qquad j > K + E \tag{2.138}$$

where the symbols $\sum_b$ and $\sum_c$ indicate a summation over all the corresponding streams, and new component variables are defined as:

$$\mathbb{y}_{jb} = y_{jb} - \sum_k \sigma_{kj} y_{kb}; \qquad j > K + E \tag{2.139}$$

$$\mathbb{F}_{jc} = F_{jc} - \sum_k \sigma_{kj} F_{kc}; \qquad j > K + E \tag{2.140}$$

The practical use of Eqs. (2.138)–(2.140) will be discussed next. Although they could also be applied to the species A_e, no benefit would arise. Therefore, the use of the primary balances (2.91) for them will be assumed hereafter.

By considering $\mathbb{N}_{jI}$, $\sum_c \mathbb{F}_{jc}$, and F_b $(1 \leq b \leq B)$ as *source terms* and taking into account that the $\mathbb{y}_{jb}$ (Eq. 2.139) are known constants, Eq. (2.138) can be decomposed by assigning a specific variable to each of the source terms, as follows:

$$\frac{dn_I}{dt} + \frac{n_I}{\theta} = 0; \qquad n_I(0) = 1 \tag{2.141}$$

$$\frac{dN_b}{dt} + \frac{N_b}{\theta} = F_b; \qquad N_b(0) = 0; \qquad b = 1, \ldots, B \tag{2.142}$$

$$\frac{d\mathbb{N}_{jc}}{dt} + \frac{\mathbb{N}_{jc}}{\theta} = \sum_c \mathbb{F}_{jc}; \qquad \mathbb{N}_{jc}(0) = 0; \qquad j > K + E \tag{2.143}$$

[*Note*: the solution of Eq. (2.141) can be expressed as $n_I = \exp(-\int_0^t \frac{1}{\theta} dt) = E^{-1}(t)$].

Instead of using the total mass balance for determining M (Eq. 2.87), $M = \sum_j m_j N_j$ can be employed, and the factor $(1/\theta) = W/M$) in Eqs. (2.141)–(2.143) can therefore be practically expressed as:

$$(1/\theta) = W/(\sum_j m_j N_j) \tag{2.144}$$

Assuming the values of n_I, N_b and $\mathbb{N}_{jc}$ have been determined, $\mathbb{N}_j(t)$ is evaluated from:

$$\mathbb{N}_j(t) = \mathbb{N}_{jI} n_I(t) + \sum_b \mathbb{y}_{jb} N_b(t) + \mathbb{N}_{jc}(t); \qquad j > K + E \tag{2.145}$$

The present formulation is completed with Eq. (2.91) for the species A_k and A_e. Three possible situations using Eqs. (2.141)–(2.144) are discussed next.

(a) **The control actions strictly contain only the species A_e and $W = 0$ (no homogeneous extraction).**

Under these conditions $1/\theta = 0$ and $\sum_c \mathbb{F}_{jc} = 0$ $(j > K + E)$ and the solution of Eqs. (2.141)–(2.143) is elementary:

- from Eq. (2.141), $n_I = 1$
- from Eq. (2.142), $N_b = \int_0^t F_b dt$
- from Eq. (2.143), $\mathbb{N}_{jc} = 0$

Then, Eq. (2.145) result in:

$$\mathbb{N}_j(t) = \mathbb{N}_{jI} + \int_0^t (\sum_b \mathbb{y}_{jb} F_b) dt; \qquad j > K + E \tag{2.146}$$

The values $\mathbb{N}_j$ can thus be evaluated in advance, independently of the system evolution. As defined in Sect. 2.2, the variables $\mathbb{N}_j$ behave as *reaction invariants* in the present situation.

The $(K + E)$ governing Eq. (2.91) of species A_k and A_e can subsequently be solved, having available values of N_j from Eqs. (2.137) and (2.146). It is noted that this is the case already considered at the end of Sect. 2.8.1.

(b) **The control actions strictly contain only the species** A_e, **but** $W \neq 0$.

As in the situation **(a)**, $\mathbb{N}_{jc} = 0$, as $\sum_c \mathbb{F}_{jc} = 0$. However, since $1/\theta \neq 0$ in this case, the $(1 + B)$ equations (2.141) and (2.142) are coupled with the rest of the $(K + E)$ conservation equations for the A_k and A_e species, as follows from Eq. (2.144). As a result, $(K + E + 1 + B)$ equations must be solved simultaneously, with the assistance of:

$$\mathbb{N}_j = \mathbb{N}_{jI} n_I + \sum_b \mathbb{y}_{jb} N_b; \qquad j > K + E \tag{2.147}$$

which follows from Eq. (2.145), and N_j from Eq. (2.137).

Therefore, the described procedure will be advantageous, as compared to the simultaneous solution of the S original balances (2.91), as long as $S - K > 1 + B + E$.

(c) **General situation: species** A_k **and** A_j $(j > K + E)$ **are (marginally) present in the control-streams and a homogeneous extraction exists** $(W \neq 0)$.

In this case, $\sum_c \mathbb{F}_{jc} \neq 0$ in Eq. (2.143) and also $1/\theta \neq 0$. Hence, Eqs. (2.141–2.144) and the conservation Eq. (2.91) for the A_k and A_e species, in a total number of $D = S + 1 + B$ equations must be solved simultaneously. Then, $1 + B$ additional equations would have to be solved, with respect to directly solving the original system of $D = S$ conservation equations for all species. However, it can be realised that the variables $\mathbb{N}_{jc}$ $(j > K + E)$ defined by Eq. (2.143) will take very small values, because of the marginal fluxes $\sum_c \mathbb{F}_{jc}$. Consequently, the impact of the $\mathbb{N}_{jc}$ in Eq. (2.145) and hence on the overall behaviour of the system will be weak. This fact can be exploited in the resolution procedure. To appreciate this possibility, it is first convenient to outline the kind of strategy needed for the numerical solution of a set of ordinary differential equations (ODEs) governing the behaviour of a reaction system.

Since the rates r_j are frequently non-linear and significantly different for different species, the system of coupled ODEs is usually *stiff*. Consequently, implicit integration methods are needed to efficiently and stably obtain the solution at a certain time t from earlier available information. Whichever the chosen implicit method, the ODEs should be discretised, given rise to a system of non-linear algebraic equations that must be first linearised (e.g., by a method of Newton–Raphson's type). A set of coupled linear equation thus arises that should be recurrently solved, as their coefficients still depend on the unknowns, until convergence is achieved. For an ODE set of D coupled equations each instance of linear resolution implies a number of elementary operations proportional to D^3.

In the present case, if the Equations (2.143), in number $S_{NE} = S - (K + E)$, are removed from the coupled equation set, D will decrease from $D = S + 1 + B$ to $D = K + E + 1 + B$. The difference $[S - (K + E)]$ can significantly reduce the computational cost of each instance of linear resolution. It is recalled that the reduced coupled set will comprise Eq. (2.91) for N_k, N_e, and Eqs. (2.141) and (2.142) for n_I and N_b. Then, for each instance of linear resolution the needed values of variables $\mathbb{N}_{jc}$ can be assumed given by suitable estimations. Once such an instance is performed, updated values of N_k, N_e, n_I and N_b will be available. With them, updated values of $\sum_c \mathbb{F}_{jc}$ and $(1/\theta)$ can also be evaluated, which in turn can be employed to update the $\mathbb{N}_{jc}$ from Eq. (2.143). These last evaluations will be made with a very low computational effort, as Eqs. (2.143) with assumed values of $\sum_c \mathbb{F}_{jc}$ and $(1/\theta)$ become linear and uncoupled from each other. The updated values $\mathbb{N}_{jc}$ are then used to perform a new instance of linear resolution and update again the values of N_k, N_e, n_I and N_b. The procedure is repeated up to final convergence of all variables.

The outlined procedure, involving the solution of $(K + E + 1 + B)$ coupled equations, can be compared to the direct solution of all $S = K + E + S_{NE}$ original coupled balances (Eq. 2.91). Hence, the larger the number S_{NE} of component species not engaged in control actions, the more efficient the outlined procedure will be. It is recalled that in actual practice the number of either pre-set or control-streams will not be larger than 1 or 2.

Finally, it should be noted that a SBR will most frequently operate without the homogeneous stream and, as seen in case **(b)** above, the variables n_I and N_b can be evaluated in advance as $n_I = 1$, $N_b = \int_0^t F_b dt$, i.e., without intervening in any iterative process. Then, the solution will imply solving only a set of $K + E$ coupled equations.

References

1. Ung, S., & Doherty, M. F. (1995). Vapor–liquid phase equilibrium in systems with multiple chemical reactions. *Chemical Engineering Science, 50*.
2. Ung, S., & Doherty, M. F. (1995). Synthesis of reactive distillation systems with multiple equilibrium chemical reactions. *Industrial & Engineering Chemistry Research, 34*.
3. Gadewar, S. B., Schembecker, G., & Doherty, M. F. (2005). Selection of reference components in reaction invariants. *Chemical Engineering Science, 60*.
4. Gillespie, K. I., & Solomons, C. (1960). Molon-a new concentration unit. *Journal of Chemical Education, 37*.
5. Prausnitz, J. M., Lichtenthaler, R. N., & Gomes de Azevedo, E. (1999). *Molecular thermodynamics of fluid-phase equilibria* (3rd ed.). Prentice-Hall.
6. Bird, R. B., Steward, W. E., Lightfoot, E. N., & Klingenberg, D. J. (2015). *Introductory transport phenomena*. Wiley.
7. Chapman, W. G., & Fouad, W. A. (2022). Beyond Flory–Huggins: Activity coefficients from perturbation theory for polar, polarizable, and associating solvents to polymers. *Industrial & Engineering Chemistry Research, 61*.

Use of the Stoichiometric Relationships in Complex Reaction Systems—Reaction Invariants

3

3.1 Introduction

It was discussed in Chap. 2 the use of the stoichiometric relationships between the net production rates r_j in reaction systems described by relatively simple models (ideal flow models), as employed for the introduction of Chemical Reaction Engineering (CRE) principles. With the only exception of the *semi-batch* reactor (Sect. 2.8 of Chap. 2), the relationships between the net production rates r_j led to equivalent relationships between the changes in the amount of the reacting species. It was also mentioned in Sect. 2.2 of Chap. 2 that such relationships can be alternatively expressed in terms of suitable combinations between the amounts of the reacting species at any time or position in the reactor, called *reaction invariants* (e.g., $\mathbb{F}_j$, Ψ_j, Ω_j), although their use for the simple reaction systems in Chap. 2 was not necessary.

The application of the stoichiometric relationships between the rates r_j for more complex models describing the behaviour of continuous-flow reaction systems is undertaken in this chapter. The motivation is the same as that in Chap. 2: the possibility to facilitate or reduce the computational load for the resolution of the conservation equations, and the identification of the minimum number of variables that allow setting the state of the system at a given time and spatial position. To this end, it will become apparent that the concept of reaction invariants is most relevant. Nonetheless, it is advanced that the models representing the undertaken reaction systems do not always admit the identification of "true" reaction invariants.

Reaction invariants have been employed in the literature for different purposes, as for the design, optimisation or control of reaction units, identification of the stoichiometry or kinetics from experimental data or to the analysis and representation of the evolution of reaction systems. References and a brief account of such applications can be found in the work of Santa Cruz et al. [1].

G. F. Barreto and C. D. Luzi, *Applying Multiple-Reaction Stoichiometry to Chemical Reactor Modelling*, Synthesis Lectures on Chemical Engineering and Biochemical Engineering, https://doi.org/10.1007/978-3-031-42375-8_3

The use of reaction invariants involves the decomposition of the set of conservation equations of the species present in a reaction system—as it will become evident through this chapter—in a subset corresponding to the conservation of the key species, usually named *reaction variants*, and another subset of conservation equations for the reaction invariants. Within a more general frame, the equations from a model representing the behaviour of a physical or chemical system can be decomposed in different ways, according to the purpose aimed with such model (see e.g., Fjeld [2]). In this sense, and of specific concern for chemical reaction systems, alternative decomposition strategies to that of reaction variants and invariants have been proposed, as for example in the works of Rodrigues et al. [3] and Billeter et al. [4] (and additional references therein), where the decomposition pursues to isolate the effects of input/extraction streams, mass exchange in heterogeneous systems or convection/diffusion transport mechanisms.

It is discussed in Sect. 3.2 that expressing the conservation equations of the chemical species in terms of their mass fractions is a suitable choice for identifying reaction invariants. Section 3.3 is devoted to the definition of combinations of the amount of the species here named *component variables,* and the conditions required for those variables to behave as reaction invariants.

For the following reaction systems that will be considered in this chapter, the component variables Ω_j, based on mass fractions, will be the first option to analyse if they can behave as reaction invariants:

(a) Tubular reactors described by the axial dispersion model
(b) Reaction systems with diffusive transport
(c) Homogeneous reactors with turbulent flow in more than one spatial coordinate
(d) Catalytic packed-bed reactors in more than one spatial coordinate.

The cases (a)–(d) will be treated consecutively in Sects. 3.4–3.7. The case (b) differs from the others because it involves pure diffusive transport. It is *per se* of practical importance, but the underlying reason for being here included is that it shows relevant aspects in close relation to systems (c) and (d). It is noted that the models presented for systems (a), (c) and (d) include effects of mass dispersion, transient behaviour, and the possibility of mass transfer through the solid boundaries.

Section 3.8 is dedicated to identifying conditions leading to relate temperature changes with composition changes when systems (a), (c) and (d) are operated adiabatically. In Sect. 3.9, it is briefly discussed reasons why other important reaction systems other than (a), (b) and (c) impose severe limitations for the worthy use of the component variables Ω_j. The main conclusions of the chapter are summarised in Sect. 3.10.

Finally, it should be noticed that all sections of the chapter are strongly linked. In particular, the discussion in Sect. 3.4 (axial dispersion model) is frequently recalled in the following sections.

3.2 The Choice of Mass Fractions to Express Conservation Equations

To compact the mathematical expressions in this chapter, the bounds in the summation symbols will be implicitly associated with the summation index, as already made in Chap. 2. The examples frequently used are:

$\sum_i \equiv \sum_{i=1}^R$: summation over some property associated with the R chemical reactions taking place;

$\sum_j \equiv \sum_{j=1}^S$: summation over some property of the S chemical species present in the system;

$\sum_{j>K} \equiv \sum_{j=K+1}^S$: summation over some property of the $(S-K)$ component species;

$\sum_k \equiv \sum_{k=1}^K$: summation over some property of the K key species (index k' is used occasionally for key species).

Similar notation will be used and explained for other occasional grouping of the species involved in the summation.

Flow reaction systems undertaken in this chapter involve effects like velocity distribution, diffusion or "disperse" transport of the species (in addition to convection) and transient conditions. These features call for the formulation of the species conservation equations in terms of some measure of the species concentrations (intensive variables), being the use of extensive variables (e.g., molar flow rates) normally impractical, at variance with the simple systems treated in Chap. 2. As a consequence, the *relative variables* employed in Chap. 2 are not useful either.

Therefore, the first step to be developed in the present chapter is to choose a measure of the concentration of the chemical species suitable to stoichiometrically relate the changes of the component-species concentrations with those of the key species. To that end, some of the conclusions reached in Sect. 2.2 of Chap. 2 are conveniently recalled, considering specifically an ideal plug-flow tubular reactor in steady state (SS). In general, molar concentrations were not found suitable to establish the desired relationships, while the use of molar fractions involves non-linear relationships, which would lead to very complex expressions for the conservation expressions. Instead, mass fractions ω_j or moles per unit mass of the mixture, $\psi_j = \omega_j/m_j$, were shown to be equally suitable. It was also discussed in Sect. 2.2 that the use of ω_j requires to restate the reaction stoichiometry in mass terms to keep the conservation equation as simple as possible, while the quantities ψ_j allow keeping the stoichiometry in the usual molar terms. Mass fractions will be employed throughout this chapter, in attention to their extensive use in the literature (e.g., Bird et al. [5]). Nonetheless, the whole formulation presented here can be based on ψ_j variables by simply replacing ω_j by $m_j\psi_j$, an operation that automatically restates the stoichiometry in molar terms.

In mass terms, the stoichiometry of a generic reaction is expressed as:

$$\mathcal{R}_i = \sum_j \hat{\nu}_{ij} A_j = 0, \tag{3.1a}$$

where the mass coefficients $\hat{\nu}_{ij}$ are related to the molar coefficients ν_{ij} by

$$\hat{\nu}_{ij} = m_j \nu_{ij} \tag{3.1b}$$

It is satisfied that:

$$\sum_j \hat{\nu}_{ij} = 0, \tag{3.1c}$$

Equation (3.1c) is a consequence of total mass conservation in a chemical reaction.

The net production rates of a species A_j, $\hat{r}_j$ (mass of A_j produced by unit time and unit volume) is given by:

$$\hat{r}_j = m_j r_j \tag{3.1d}$$

In term of the reaction rates r_i

$$\hat{r}_j = \sum_i \hat{\nu}_{ij} r_i \tag{3.1e}$$

For the normalised canonical reactions $\mathcal{R}_k = \sum_j \hat{\sigma}_{kj} A_j = 0$, $\hat{\sigma}_{kk'} = 1$ $(k = k')$, $\hat{\sigma}_{kk'} = 0$ $(k \neq k')$, and for the component species,

$$\hat{\sigma}_{kj} = \frac{m_j}{m_k} \sigma_{kj}; \qquad j > K \tag{3.2a}$$

Note that in this way $\hat{\sigma}_{kj}$ expresses mass of A_j per unit mass of A_k. Also,

$$1 + \sum_{j>K} \hat{\sigma}_{kj} = 0 \tag{3.2b}$$

The stoichiometric relations stated in Chap. 1, $r_j = \sum_k \sigma_{kj} r_k$, are re-rewritten as

$$\hat{r}_j = \sum_k \hat{\sigma}_{kj} \hat{r}_k; \qquad j > K \tag{3.3}$$

Since the kinetic expressions of the rates $\hat{r}_j$ are usually expressed as functions of molar concentrations C_j, it is recalled that:

$$C_j = \rho \omega_j / m_j, \tag{3.4}$$

3.3 Definition of Component Variables and Reaction Invariants

As noted in the Introduction, the use of reaction invariants involves the decomposition of the set of conservation equations of the species present in a reaction system. An ideal plug-flow tubular reactor in steady state (SS) is considered as an elementary example of such decomposition. Conservation equations and boundary conditions in terms of mass fractions, discriminated between key and component species, are:

$$G\frac{d\omega_k}{dz} = \hat{r}_k; \qquad k = 1, \ldots, K \tag{3.5a}$$

$$\omega_k(0) = \omega_{k0} \tag{3.5b}$$

$$G\frac{d\omega_j}{dz} = \sum_k \hat{\sigma}_{kj}\hat{r}_k; \qquad j > K \tag{3.6a}$$

$$\omega_j(0) = \omega_{j0} \tag{3.6b}$$

where z is the axial variable and G is the mass velocity, uniform along z.

The appearance of the rates $\hat{r}_k$ in Eqs. (3.6a, b) can be eliminated if each of them is replaced by its expression in Eqs. (3.5a, b). The results can be written as:

$$G\frac{d\Omega_j}{dz} = 0; \qquad j > K \tag{3.7a}$$

$$\Omega_j(0) = \Omega_{j0} \tag{3.7b}$$

where the introduced variables Ω_j are defined as:

$$\Omega_j = \omega_j - \sum_k \hat{\sigma}_{kj}\omega_k; \qquad j > K \tag{3.8}$$

The operation performed to obtain Eqs. (3.7a, b) can be symbolised as $[(3.6)_{j>K} - \sum_k (3.5)_k \times \hat{\sigma}_{kj}]$. This kind of combinations between equations will be extensively used throughout this chapter. As in the example, such an operation leads to decompose the set of species balances and their boundary/initial conditions in two subsets: the original one for the individual key species and the other one for the combined variables (defined further below as *component variables*, Ω_j in the example), in which all reaction rates have been eliminated.

It follows from Eq. (3.7a, b) that Ω_j will remain constant along the reactor and equal to the value at the entry position $z = 0$:

$$\Omega_j = \Omega_{j0}; \qquad j > K \tag{3.9}$$

Then, to evaluate the composition throughout the reactor, it is only necessary to integrate the subset of the differential Eqs. (3.5a, b) for the ω_k, with local values of ω_j $(j > K)$ (needed to evaluate the rates $\hat{r}_k$) given from the definitions (3.8) as:

$$\omega_j = \Omega_j + \sum_k \hat{\sigma}_{kj}\omega_k; \qquad j > K \tag{3.10a}$$

Alternatively, by writing from Eq. (3.9) $\Omega_j = \Omega_{j0} = \omega_{j0} - \sum_k \hat{\sigma}_{kj}\omega_{k0}$:

$$\omega_j = \omega_{j0} + \sum_k \hat{\sigma}_{kj}(\omega_k - \omega_{k0}); \qquad j > K \tag{3.10b}$$

Explicit stoichiometric relationships between the amounts of key and component species, similar to Eq. (3.10b), were shown to apply in the reaction systems studied in Chap. 2. This feature made unnecessary the use of variables like Ω_j and their elementary conservation equations (i.e., Eqs. 3.7a, b) in order to reduce the number of coupled conservation equations that should be solved, from S to K. Instead, to analyse if such a reduction is feasible for the reaction systems treated in this chapter, the introduction of variables like Ω_j for the component species will be most convenient.

Some general properties of the variables Ω_j follow from the definition in Eq. (3.8). If it is applied for any key species, the result will be meaningless, $\Omega_k = 0$. Instead, it is stressed that Eqs. (3.7a, b) should be extended to chemically inert species, for which Eq. (3.8) states that $\Omega_j = \omega_j$, since $\hat{\sigma}_{kj} = 0$. By adding Eq. (3.8) on $j > K$ (including inert species), it is verified by virtue of Eq. (3.2b) that for any reaction system:

$$\sum_{j>K} \Omega_j = 1 \tag{3.11}$$

Although variables Ω_j are chiefly used throughout this chapter, it will be useful to consider a generic definition of *component variables* and their relationship with the concept of reactions invariants.

Component Variables:

Consider a variable V_j proportional or just equal to a given measure (either intensive or extensive) of the mass instantaneously present of a component species A_j at a given time and position of a reaction system. The proportionality coefficient, identified as κ_j, is chosen by convenience, as a known value of a property of species A_j, a property of the mixture, an operating condition or some combination of the formers.

The associated *component variable* $\mathbb{V}_j$ is defined as:

$$\mathbb{V}_j = V_j - \sum_k \hat{\sigma}_{kj} V_k; \qquad j > K \tag{3.12a}$$

A variable $\mathbb{V}_j$ is thus a stoichiometry-based combination of the variable V_j of a component species ($j > K$) with the key-species variables V_k.

It is noted that for any chemically inert species, $\mathbb{V}_j = V_j$.

The variables Ω_j defined in Eq. (3.8) are the component variables of the mass fractions ω_j, i.e., $V_j = \omega_j$, with a common proportionality coefficient, $\kappa_j = \kappa = 1$.

The meaning of $\mathbb{V}_j$ can be related to the hypothetical event of complete consumption of all key species A_k existing at a given position and time. In particular, when a common proportionality coefficient is chosen ($\kappa_j = \kappa$), $\mathbb{V}_j$ will account for the total mass of A_j that would remain after such an event.

By adding Eq. (3.12a) for all $j > K$ and recalling Eq. (3.2b), it is obtained:

$$\sum_{j>K} \mathbb{V}_j = \sum_j V_j = V_T \tag{3.12b}$$

If $\kappa_j = \kappa$ is chosen, V_T will account for the total mass present at the given position and time. For example, $\sum_{j>K} \Omega_j = \sum_j \omega_j = 1$.

It should be noted that depending on the choice of the component species, values $\mathbb{V}_j < 0$ and/or $\mathbb{V}_j > V_T$ can result for some species. Nonetheless, in particular when $\kappa_j = \kappa$, it can be expected that there will be some choice of the set of component species allowing that $0 \leq \mathbb{V}_j \leq V_T \;\; \forall \;\; j > K$.

Component variables in terms of molar amounts can also be used. Starting from given variables V_j and the definition of $\mathbb{V}_j$ in Eq. (3.12a), it can be written:

$$\mathbb{V}_j / m_j = V_j / m_j - \sum_k \hat{\sigma}_{kj} V_k / m_j; \qquad j > K$$

Then, by defining $V_j / m_j = V'_j$ (for any j) and $\mathbb{V}'_j = \mathbb{V}_j / m_j$ (for $j > K$) and by considering Eq. (3.2a),

$$\mathbb{V}'_j = V'_j - \sum_k \sigma_{kj} V'_k; \qquad j > K \tag{3.12c}$$

Thus, V'_j and their associated component variables $\mathbb{V}'_j$ are related to moles of A_j, with molar stoichiometric coefficients σ_{kj} involved.

Variables V'_j relevant in the context of this chapter are $\psi_j = \omega_j / m_j$ with the associate component variables $\Psi_j = \psi_j - \sum_k \sigma_{kj} \psi_k$. As already mentioned, ψ_j can be equivalently used instead of ω_j, and consequently Ψ_j instead of Ω_j.

Reaction Invariants:

Assume that it is possible to decompose the species balances in a subset of governing equations for the amount of the key species (K equations) and another subset of governing equations for some component variables $\mathbb{V}_j$ ($S-K$ equations). Then, if the variables $\mathbb{V}_j$ keep independency of any effect of the chemical reactions, even under certain assumptions, it is said that the $\mathbb{V}_j$ are *reaction invariants* for the model applied and under the (eventual) assumptions. Accordingly, it is emphasised that the existence of reaction invariants is subject to the model employed rather than to the actual reaction system.

An immediate consequence of the above definition for reaction invariants is that the governing subset of equations for variables $\mathbb{V}_j$ can be independently solved. Once the fields of the $\mathbb{V}_j$ are thus evaluated, it can be written from Eq. (3.12a) for each position and time:

$$V_j = \mathbb{V}_j + \sum_k \hat{\sigma}_{kj} V_k; \qquad j > K \tag{3.13}$$

Recalling that each V_j is related to the amount of A_j, Eq. (3.13) is used to solve the equation subset for the key species. In this way, the uncoupled solution of both subsets will reduce the overall computational demand. In addition, the solution of the subset for $\mathbb{V}_j$ will be relatively easy to carry out, as reaction rates are not involved. It can be also concluded that the system composition becomes just established by the resulting values of V_k.

For the steady-state ideal plug-flow tubular reactor, Eqs. (3.7a, b) reveal that the variables Ω_j behave as reaction invariants, without no further assumption, i.e., for any temperature and pressure distributions inside the reactor and for any reaction set.

3.4 Tubular Reactors Described by the Axial Dispersion Model

The use of the component variables Ω_j for tubular reactors described by the Axial Dispersion Model (ADM) is discussed in this section. In addition to convective transport, the basic form of the ADM recognises the axial diffusion/dispersion transport of the species along the reactor, which is basically quantified by an effective axial dispersion coefficient D, common for all species and spatially uniform. A more general description will be used here by regarding that dispersion coefficients can be specific for each species and that they can vary axially. In addition, transient state conditions and the possibility for mass transfer through the reactor wall are considered. The latter feature will arise, for instance, when the reactor is laterally bounded by a permeable membrane.

Although the study in this section is focussed on homogeneous reactors, the ADM can also be applied for any or both flowing phases of a heterogeneous-flow reactor, as in reactive absorption processes. The formulation given here will apply and the discussion in this section will therefore still be valid for any of the streams, provided that some distinct features are taken into account. The hold-up of each phase and the interfacial surface for mass exchange should be evaluated according to the operating conditions, and dispersion coefficients will be different from those for a homogeneous stream. Furthermore, the

ADM can also be used for catalytic packed-bed reactors, but its formulation will arise as a particular case of the more general model treated in Sect. 3.7.

The axial dispersion mass-fluxes j_j will be formulated by a Fickian type of equation:

$$j_j = -\rho D_j \partial\omega_j/\partial z, \tag{3.14}$$

where D_j is the axial dispersion coefficient of A_j. The dispersion fluxes j_j are defined with respect to the mass-average velocity, $u = G/\rho$. Some important aspects concerning Eq. (3.14) and coefficients D_j will be discussed next.

Values of D_j depends strongly on the fluid-dynamic regime and hence on the Reynolds number, $Re = Gd_t/\mu$ (d_t: hydraulic diameter of the tube). In laminar regime ($Re < 2100$, for circular tubes), D_j is expressed by:

$$\text{(laminar regime)} \qquad D_j = \mathfrak{D}_j + D_{j,Taylor} \tag{3.15a}$$

$$D_{j,Taylor} = (ud_t)^2/(192\mathfrak{D}_j) \tag{3.15b}$$

where $\mathfrak{D}_j$ is the molecular diffusivity of A_j in the mixture and $D_{j,Taylor}$ is the *Taylor dispersion coefficient*, which is caused by the transversal profile of axial velocities. $D_{j,Taylor}$ is normally the major contribution in Eq. (3.15a). The coefficient 192 in Eq. (3.15b) should be modified for tubes with cross-section shapes different than the circular one (see e.g., Levenspiel [6]).

The molecular diffusivity $\mathfrak{D}_j$ is unambiguously defined for binary mixtures, when $\mathfrak{D}_j$ is the binary diffusion coefficient $\mathfrak{D}_{j,i}$ for the species pair. Similarly, for diluted species in a solvent A_s, the molecular diffusivity $\mathfrak{D}_j$ of each species can be considered as a "binary" diffusion coefficient between A_j and the solvent ($\mathfrak{D}_j \approx \mathfrak{D}_{j,s}$). In a general case, an approximation is needed for the evaluation of $\mathfrak{D}_j$. It is considered here a Wilke's type approximation [7], as given in Sect. 12.7.4 of the text by Kee et al. [8]:

$$\frac{1-\omega_j}{\mathfrak{D}_j} = \sum_{i \neq j} \frac{y_i(1-\omega_j) + \omega_i y_j}{\mathfrak{D}_{j,i}}, \tag{3.16}$$

where the summation index i extends over all species other than A_j. An additional contribution to dispersion arises in turbulent regime, due to fluctuating components of the velocity causing turbulent mixing. Instead, the Taylor contribution to dispersion decreases because the transverse profile of axial velocity becomes less pronounced. The empirical expression of Wen and Fan (see Sect. 19 in Green and Perry [9]) can be employed to estimate an effective dispersion coefficient D at values $Re > 2000$:

$$\text{(turbulent regime)} \qquad \rho D = d_t G\left(\frac{3 \times 10^7}{Re^{2.1}} + \frac{1.35}{Re^{1/8}}\right) \tag{3.17}$$

Expression (3.17) provides a common value of D for all species, $D_j = D$. Levenspiel [10] pointed out that the experimental values of D up to around $Re = 10000$ are considerable disperse, a fact that partially can arise for some effect of the Schmidt number of the individual species $Sc_j = \mu/(\rho\mathfrak{D}_j)$. It follows that values of D from Eq. (3.17) in the range $2000 < Re < 10000$ will have certain level of uncertainty.

It should be noted that any approximation for the multi-component fluxes j_j based on a Fickian type of expression, like Eq. (3.14) or the equivalent expression just for the molecular fluxes, i.e., $j_j = -\rho\mathfrak{D}_j\partial\omega_j/\partial z$, will not satisfy the consistence condition $\sum_j j_j = 0$, when the coefficients D_j, or $\mathfrak{D}_j$, are different. Nonetheless, Fickian fluxes are frequently used in flow system, even for rigorous simulations, as those based on computational fluid dynamics (CFD) (see e.g., Jurtz et al. [11]), sustained by the assumption that dispersion or diffusion transport is of secondary importance with respect to convection. For simplicity reasons, the governing equations of the individual species will be presented in this chapter with Fickian fluxes (e.g., Eq. 3.14 for the ADM). For the main purpose of exploring the behaviour of the Ω_j, such a simplification will not bring any consequence. In actual calculations, however, it would be advisable to introduce a proper scheme to evaluate the dispersion/diffusion fluxes in a way to satisfy the consistence condition. An alternative to that purpose suggested by Kee et al. [8] is presented in Appendix 1.

It will be considered in this chapter that the different reaction systems can show temperature variations. However, for the purpose of analysing the use of component variables and their possible behaviour as reaction invariants, it will not be necessary to write the energy conservation equation, with the exception of Sect. 3.8 where adiabatic reactors are discussed. A similar consideration applies for the momentum balance needed to evaluate the pressure distribution.

The mass conservation equations for the ADM under the conditions mentioned above will be presented next. The mass conservation and the condition at the reactor entry, $z = 0$, are expressed as:

$$\frac{\partial \rho}{\partial t} + \frac{\partial G}{\partial z} + a_w G_w = 0; \tag{3.18a}$$

$$G(t, 0) = G_0(t); \tag{3.18b}$$

where $G(t, z)$ is the mass velocity, the input value $G_0(t)$ is a given function of time, (constant in the SS), a_w is the ratio between the perimeter and cross-section area of the tube (inverse of the hydraulic radius), $G_w = \sum_j G_{jw}$ is the total mass flux through the wall and G_{jw} the mass flux of species A_j (positive if out of the reactor).

It should be noted that the evaluation of the fluxes G_{jw}, for example through a membrane, will have to be done in practice by considering the mass transfer steps in the tube side, through the membrane itself and in the external carrier. For the present purpose, the details of such a procedure are not essential. It is sufficient to point out that the fluxes

G_{jw} at each axial position will depend on the local composition in the tube and on the membrane selectivity.

By expressing the mass flux of species A_j as $G_j = \omega_j G + j_j = \omega_j G - \rho D_j \partial\omega_j/\partial z$, the conservation equation inside the reactor, initial condition, and Danckwerts' boundary conditions are:

$$\frac{\partial}{\partial t}(\rho\omega_j) + \frac{\partial}{\partial z}(\omega_j G) - \frac{\partial}{\partial z}\left(\rho D_j \frac{\partial \omega_j}{\partial z}\right) + a_w G_{jw} = \hat{r}_j; \tag{3.19a}$$

$$\begin{aligned} &\omega_j(0, z) = \omega_{jI}(z); \\ &\left[G_0(\omega_j - \omega_{j0}) - \rho D_j \frac{\partial \omega_j}{\partial z}\right]_{z=0} = 0; \\ &\left(\frac{\partial \omega_j}{\partial z}\right)_{z=L} = 0; \end{aligned} \tag{3.19b}$$

where L is the tube length and $\omega_{j0}(t)$ is the known value of ω_j in the input stream. Using the product rule for $\partial(\omega_j G)/\partial z$, Eq. (3.18a) for $\partial G/\partial z$ and discriminating between key and component species, it is obtained:

For the key species, $k = 1, \ldots, K$:

$$\rho\frac{\partial \omega_k}{\partial t} + G\frac{\partial \omega_k}{\partial z} - \frac{\partial}{\partial z}\left(\rho D_k \frac{\partial \omega_k}{\partial z}\right) + a_w(G_{kw} - \omega_k G_w) = \hat{r}_k; \tag{3.20a}$$

$$\begin{aligned} &\omega_k(0, z) = \omega_{kI}(z); \\ &\left[G_0(\omega_k - \omega_{k0}) - \rho D_k \frac{\partial \omega_k}{\partial z}\right]_{z=0} = 0; \\ &\left(\frac{\partial \omega_k}{\partial z}\right)_{z=L} = 0; \end{aligned} \tag{3.20b}$$

For the component species, $j > K$:

$$\rho\frac{\partial \omega_j}{\partial t} + G\frac{\partial \omega_j}{\partial z} - \frac{\partial}{\partial z}\left(\rho D_j \frac{\partial \omega_j}{\partial z}\right) + a_w\left(G_{jw} - \omega_j G_w\right) = \sum_k \hat{\sigma}_{kj}\hat{r}_k \tag{3.21a}$$

$$\begin{aligned} &\omega_j(0, z) = \omega_{jI}(z); \\ &\left[G_0(\omega_j - \omega_{j0}) - \rho D_j \frac{\partial \omega_j}{\partial z}\right]_{z=0} = 0; \\ &\left(\frac{\partial \omega_j}{\partial z}\right)_{z=L} = 0; \end{aligned} \tag{3.21b}$$

By making the operation $[(3.21a, b)_{j>K} - \sum_k (3.20a, b)_k \times \hat{\sigma}_{kj}]$ between these equations and recalling the definition of the component variables Ω_j (Eq. 3.8), it is obtained the subset of equations, initial and boundary conditions for Ω_j:

$$\begin{aligned} &\rho\frac{\partial \Omega_j}{\partial t} + G\frac{\partial \Omega_j}{\partial z} - \frac{\partial}{\partial z}\left(\rho D_j \frac{\partial \Omega_j}{\partial z}\right) = \\ &\frac{\partial}{\partial z}\left[\rho \sum_k \hat{\sigma}_{kj}(D_j - D_k)\frac{\partial \omega_k}{\partial z}\right] - a_w(\mathbb{G}_{jw} - \Omega_j G_w); \qquad j > K \end{aligned} \tag{3.22a}$$

$$\Omega_j(0,z) = \Omega_{jI}(z);$$
$$G_0(\Omega_j - \Omega_{j0})_{z=0} - \left(\rho D_j \frac{\partial \Omega_j}{\partial z}\right)_{z=0} = \left[\rho \sum_k \hat{\sigma}_{kj}(D_j - D_k)\frac{\partial \omega_k}{\partial z}\right]_{z=0}; \qquad (3.22b)$$
$$\left(\frac{\partial \Omega_j}{\partial z}\right)_{z=L} = 0;$$

The component variables $\mathbb{G}_{jw}$ introduced in Eq. (3.22a) are defined as

$$\mathbb{G}_{jw} = G_{jw} - \sum_k \hat{\sigma}_{kj} G_{kw}; \qquad j > K \qquad (3.22c)$$

In using Eqs. (3.20a, b) for the key species and Eqs. (3.22a, b) for the component variables Ω_j, it is recalled that the fractions ω_j $(j > K)$ are given in Eq. (3.10a).

3.4.1 Limitations for the Existence of Reaction Invariants in Flow Systems

It can be stated that a sufficient condition for a given component variables set (e.g., Ω_j) to behave as reaction invariants is that the resolution of their governing equations (e.g., Eqs. 3.22a, b) does not involve any unassigned property that depends on the system state-variables (i.e., composition, temperature, pressure), since these will in turn depend on the reaction rates $\hat{r}_k$.

An inspection to Eq. (3.22a, b) reveals that their solution will depend on the terms presenting the differences $(D_j - D_k)$ and $(\mathbb{G}_{jw} - \Omega_j G_w)$, whenever they are non-zero. Differences $(D_j - D_k) \neq 0$ will introduce an effect of the derivatives $(\partial\omega_k/\partial z)$ on the solution, preventing the behaviour of the Ω_j as reaction invariants. The same conclusion is achieved for operations with interfacial mass transfer, as the terms $(\mathbb{G}_{jw} - \Omega_j G_w)$ will be normally non-zero (an exception is for the practically unusual non-selective mass transfer, $G_{jw} = \omega_j G_w$) and the dependence of the fluxes G_j on the system composition will be inherited by the Ω_j. Actually, the effects caused by both differences, $(D_j - D_k)$ and $(\mathbb{G}_{jw} - \Omega_j G_w)$, involve by different means the segregation of some species from the flowing mixture.

Specifically for the transient state, a third cause preventing the behaviour of Ω_j as reaction invariants is the dependency of density ρ and dispersion coefficients D_j on the state variables. This effect can be better explained by assuming a common dispersion coefficient $(D_j = D)$ and impermeable walls $(G_j = 0)$, thus eliminating the segregation effects. Equations (3.22a, b) become:

$$\rho\frac{\partial \Omega_j}{\partial t} + G\frac{\partial \Omega_j}{\partial z} - \frac{\partial}{\partial z}\left(\rho D \frac{\partial \Omega_j}{\partial z}\right) = 0; \qquad j > K \qquad (3.23a)$$

$$\Omega_j(0,z) = \Omega_{jI}(z);$$
$$G_0\left(\Omega_j - \Omega_{j0}\right)_{z=0} - \left(\rho D \frac{\partial \Omega_j}{\partial z}\right)_{z=0} = 0; \tag{3.23b}$$
$$\left(\frac{\partial \Omega_j}{\partial z}\right)_{z=L} = 0;$$

The total mass balance, Eqs. (3.18a, b), results:

$$\frac{\partial \rho}{\partial t} + \frac{\partial G}{\partial z} = 0; \tag{3.23c}$$

$$G(t,0) = G_0(t); \tag{3.23d}$$

It is first noted that in the SS (constant entry conditions), the temporal derivatives vanish and Eqs. (3.23a, b) are satisfied with uniform values $\Omega_j = \Omega_{j0}$, regardless of the values of ρ, D and G, and thus the Ω_j are reaction invariants.

During a transient state, for example before reaching the SS at constant input values, the initial values $\Omega_{jI}(z)$ can frequently be different to Ω_{j0}. The variables Ω_j will evolve from $\Omega_{jI}(z)$ to Ω_{j0} and thus $\partial\Omega_j/\partial t \neq 0$ in Eq. (3.23a), causing the appearance of non-zero spatial derivatives. In this way, the effect of ρ and D are introduced in the solution. Then, if the dependence of ρ and D with the state variables is considered, the variables Ω_j will not behave as reaction invariants during the transient period.

A different scenario arises if the reactor is operating in steady state and only the input temperature is changed, without modifying the input composition (and thus leaving the Ω_{j0} invariable). In such a case, during the transient period to reach the new steady state, the solution $\Omega_j = \Omega_{j0}$ will still apply, provided that $D_j = D$.

The discussed limitations will be valid in general for any model accounting for segregation effects and independent initial/input conditions Ω_{jI} and Ω_{j0}. They are summarised and listed below, as reference for the next sections.

Limitations for the behaviour of Ω_j as reaction invariants in flow systems

Variables Ω_j will not behave as reaction invariants if any of the following conditions takes place:

(A) Different values of dispersion coefficients, particularly $(D_j - D_k) \neq 0$
(B) Existence of selective interface mass transfer, causing $G_{jw} \neq \omega_j G_w$, and then $\left(\mathbb{G}_{jw} - \Omega_j G_w\right) \neq 0$
(C) For a transient state, different composition between initial condition and steady state input ($\omega_{jI} \neq \omega_{j0}$, and then $\Omega_{jI} \neq \Omega_{j0}$) and when the dependency of density ρ and/or dispersion coefficients D_j on the state variables is considered.

3.4.2 Particular Cases in the Application of the Axial Dispersion Model

Different situations of practical interest can be identified when the use of variables Ω_j can be beneficial for the solution of the governing equations, even if they do not strictly behave as reaction invariants.

3.4.2.1 Particular Case 1: Variables Ω_j Behave as Reaction Invariants

To this end, for the steady state it will be necessary to avoid the limitations (A) and (B) in Sect. 3.4.1. For the former, a common dispersion coefficient D should be assumed, which is quite reasonable in turbulent regime. Equation (3.17) or any other similar expression can be used to evaluate D. The limitation (B) is simply removed by considering impermeable walls, a case most frequently found in practice. Then, Eqs. (3.23a, b) will be applicable. As already noted, in a steady state $\Omega_j = \Omega_{j0}$ are reaction invariants. The equation subset (3.20a, b) for the key species (with $D_k = D$ and $G_{kw} = 0$) can be solved by evaluating the mass fractions of the component species as $\omega_j = \Omega_j + \sum_k \hat{\sigma}_{kj}\omega_k$ (Eq. 3.10a).

It is important to note that in steady state and with impermeable walls, at the reactor exit it will be verified that $\Omega_{jL} = \Omega_{j0}$, regardless of any dispersion effect inside the reactor. This conclusion arises by integration of Eq. (3.22a) and considering the boundary conditions (3.22b). More generally, it can be obtained from global species balances between the input and outlet pipes of the reactor housing, where pure convective flow with uniform composition can be assumed. This condition equally applies for the systems in Sects. 3.6 and 3.7 under SS and impermeable walls conditions.

For a transient state, the limitation (C) should also be removed by assuming that ρ and D remains independent of the state variables. The suitability of this approximation for a given problem should be specifically evaluated. The assumption for ρ will be acceptable in many situations of reactions in liquid phase. The dependence of D on the mass velocity G (e.g., as in Eq. 3.17) does not introduce any dependence on the state variables when ρ is constant, as from (Eq. 3.23c, d), $G = G_0(t)$.

For constant values of ρ and D, it is noted that Eqs. (3.23a, b) are linear and uncoupled. Then, they are individually solved for each Ω_j. An additional simplification arises if the initial values Ω_{jI} do not depend on z, as will happen if the initial conditions correspond to a SS, and input values change suddenly to new constant values Ω_{j0}. Noting that the linear terms of all Eqs. (3.23a, b) present the same coefficients and they only differ in the constant "source" terms Ω_{jI} and Ω_{j0}, the *superposition principle* can be applied. Let $\Omega^{1,0}(t,z)$ be the solution of Eqs. (3.23a, b) for $\Omega_{jI} = 1$ and $\Omega_{j0} = 0$, and $\Omega^{0,1}(t,z)$ the solution for $\Omega_{jI} = 0$ and $\Omega_{j0} = 1$. Then for any of the reaction invariants,

$$\Omega_j(t,z) = \Omega_{jI}\Omega^{1,0}(t,z) + \Omega_{j0}\Omega^{0,1}(t,z)$$

In this way, only two independent linear differential equations should be resolved.

3.4.2.2 A Resolution Procedure of the Governing Equations for Small Deviations of Variables Ω_j as Reaction Invariants

Consider situations in which the Ω_j do not behave as reaction invariants because any of the limitations (A), (B) and (C) in Sect. 3.2 cannot be disregarded. In such cases, the introduction of variables Ω_j does not lead to an evident benefit, as their governing equations (Eq. 3.22a, b) are implicitly coupled to those of the key species (Eqs. 3.20a, b). However, an overview of the numerical procedures to find the solution of the problem will reveal that the use of the variable Ω_j may still contribute to reduce the amount of calculations.

The discussion in this section is focussed on problems with impermeable walls (limitation B is removed), while Sect. 3.4.2.3 deals specifically with situations involving interfacial mass transfer.

For their solution, the differential equations expressing the species balances, e.g., Eqs. (3.20a, b) and (3.21a, b), should be first discretised into a set of coupled algebraic equations for the unknown variables in a certain number of spatial nodes and (for transient state problems) temporal levels. Mainly due to the frequently strong non-linearity of the rates $\hat{r}_k$ and the fast variations of the concentrations of some species, the set of algebraic equations should be solved iteratively by linearising the rates $\hat{r}_k$ in each iteration step and solving the resulting set of linear equations. The number of unknown variables and equations will be proportional to the number of species S, with a proportionality coefficient depending on the number of nodes. Then, the number of elementary operations to solve the linear system in each iteration step can be estimated as proportional to S^3, which can clearly lead to a very large computational burden to reach the final solution. In view of the strong impact of the number of species S, to gain numerical efficiency, the equation set can be split into subsets, each of them alternatively used to update the associated variables while maintaining fixed values of the remaining variables. For example, with two subsets of equal size, the total number of operations in each iteration step may be expected to be of order $2(S/2)^3 = (1/4)S^3$, rather than S^3. The success of this type of strategies (usually known as *tearing methods*) heavily depends on the choice of the subsets and their associated variables.

The concept above can be tried for the subsets (3.20a, b) and (3.22a, b) with ω_k and Ω_j as iteration variables, respectively. Provided that the component variables Ω_j depend slightly on the state variables, they will closely behave as reaction invariants and the strategy is most likely to succeed (in the limit of the Ω_j being truly reaction invariant, Eqs. 3.22a, b must be solved only once). It should be noted that updating the Ω_j for given values of ω_k will require a very modest computational effort, as Eqs. (3.22a, b) become uncoupled and linear in each Ω_j.

Some details are next itemised for a basic application of the proposed procedure.

i. For a transient state problem that requires accounting for density variations, Eqs. (3.18a, b) for the total mass conservation should be added to the subset of Eqs. (3.20a, b) for the key species. The energy and momentum equations should also be included for non-isothermal problems and when the distribution of pressure is required. In this way, the extended subset will be used to simultaneously update variables ω_k, G, T, and P.
ii. The discretisation grid for the subsets of variables (ω_k, G, T, P) and Ω_j should be conveniently the same.
iii. Given updated (fixed) values of Ω_j in all nodes, the next updating step for variables (ω_k, G, T, P) can be carried out. The values ω_j needed to evaluate the rates $\hat{r}_k$, ρ and D_j (and eventually μ) in each node are evaluated as $\omega_j = \Omega_j + \sum_k \hat{\sigma}_{kj}\omega_k$.
iv. After completing the step described in item iii, updated values of ω_k, G, ρ, D_j in each node will be available for their use in Eqs. (3.22a, b). Then, new updated values of Ω_j are evaluated.
v. In general, Values of Ω_j calculated from Eqs. (3.22a, b) will not satisfy the condition $\sum_{j>K}\Omega_j = 1$, partially because of the inaccuracies introduced by the discretisation of the differential equations and also when the method for evaluating the dispersion coefficients D_j cannot warrant the fulfilment of the condition $\sum_j j_j = 0$. In addition, the updated values of Ω_j should not lead to negative values of $\omega_j = \Omega_j + \sum_k \hat{\sigma}_{kj}\omega_k$ for the currently available values of ω_k. In Appendix 2, it is described a way to normalise the final updated values of Ω_j so that they can simultaneously satisfy $\sum_{j>K}\Omega_j = 1$ and $\omega_j \geq 0$ ($j > K$). It is noted that $\sum_{j>K}\Omega_j = 1 \iff \sum_j \omega_j = 1$, by virtue of Eq. (3.12b).

The proposed procedure can be suitable for situations discussed in the oncoming Sects. 3.4.2.3, 3.4.2.4, 3.6.2.2–3.6.2.4 and 3.7.1. It has also been satisfactorily applied in *Example 1* presented later on this chapter, in which the Ω_j deviate moderately from the behaviour as reaction invariants. It should be noted we are not aware of the use in the literature of a similar procedure. Therefore, further numerical and/or analytical studies would be necessary for assessing how close the Ω_j should approach the condition of reaction invariance to avoid risks as inefficiency or, even worse, lack of convergence.

3.4.2.3 Particular Case 2: Variables Ω_j Closely Behave as Reaction Invariants

Different values D_j are considered, but under conditions of moderate dispersion effects, which can be expected to hold for low enough values of the dimensionless group $[\rho D_j(-\hat{r}_j/\omega_j)/G^2]$ for the reactant species, for which the limiting value of 1 may be adequate. This condition will be fulfilled in many practical situations.

The walls should be impermeable ($G_{jw} = 0$) and moderate variations of ρ and D_j should take place in a transient state.

Then, although limitations (A) and (C) will prevent the Ω_j from behaving as reaction invariants, moderate deviations can be expected and, consequently, the resolution procedure in Sect. 3.4.2.2 could be successfully applied.

3.4.2.4 Particular Case 3: Selective Interfacial Mass Transfer in Steady State

The use of membranes in chemical reactors responds to the purpose of increasing the process efficiency or safety. For example, a reactant can be fed through the membrane to maintain a nearly uniform concentration, a condition that can increase the selectivity to the desired products or may be employed to moderate the rate of the heat released by the reactions. In extractive membranes, a product is typically removed from the reaction mixture to increase the conversion for equilibrium-limited reactions. In most applications, as in the previous examples, the remainder species will be marginally exchanged.

Similarly, if the ADM is applied to one of the streams in heterogeneous-flow reactors, as mentioned in Sect. 3.4, the interfacial mass transfer will also involve the transfer of only some species.

Therefore, for the discussion in this section it will be assumed that only a small number E of the species are significantly exchanged through the interfacial surface, while the exchange of the reminder species is strongly impaired.

It is assumed that the E species A_e with significant rates of mass transfer can be assigned as component species, $K+1 \leq e \leq K+E$. The remainder $(S-K-E)$ component species A_j $(j > K + E)$ and the K key species show marginal rates of mass transfer.

It will be verified that $G_w \approx G_{Ew}$, where $G_{Ew} = \sum_e G_{ew}$ (where $\sum_e$ indicates the sum over all species A_e), and $G_{jw} \approx 0$ $(j \neq e)$. In the balances of the Ω_j, the transfer terms will be, approximately, $a_w(G_{ew} - \Omega_e G_{Ew})$ for the Ω_e and $a_w(-\Omega_j G_{Ew})$ for the remainder Ω_j $(j > K + E)$. In general, all these terms will be numerically significant and their dependency on the composition will prevent the Ω_j from behaving as reaction invariants (limitation C in Sect. 3.4.1). It can be verified that the choice of species A_e as key species does not alter such a conclusion, which in essence is caused by the fact that the variables Ω_j are related to the total mass in the mixture that does not remain constant, due to the interfacial exchange. This circumstance suggests looking for component variables defined in an alternative way.

To this end, it is considered a system in SS with ideally selective transfer of species A_e ($G_w = G_{Ew}$). By integrating Eq. (3.22a) for all Ω_j $(j > K)$, considering the boundary conditions (3.22b), and Eq. (3.18a) for total mass from the entry to the exit (index L), it is obtained:

$$
\begin{aligned}
&G_L = G_0 - N_{Ew}; \\
&G_L \Omega_{jL} = G_0 \Omega_{j0}, \quad (j \neq e); \\
&G_L \Omega_{EL} = G_0 \Omega_{E0} - N_{Ew};
\end{aligned}
\tag{3.24}
$$

where Ω_E is in general defined as:

$$\Omega_E = \sum_e \Omega_e = \sum_e (\omega_e - \sum_k \hat{\sigma}_{ke}\omega_k) \tag{3.25a}$$

$N_{Ew} = \int_0^L a_w G_w dz$ is the total amount of mass of species A_e transferred through the interface by unit time and cross-section area of the tube.

From Eq. (3.24), the following relationship arises:

$$\frac{\Omega_{jL}}{1 - \Omega_{EL}} = \frac{\Omega_{j0}}{1 - \Omega_{E0}}; \qquad j > K + E \tag{3.25b}$$

Then, the *modified component variables* defined for A_j, $j > K + E$, as:

$$\Omega_j^* = \frac{\Omega_j}{1 - \Omega_E}; \qquad j > K + E \tag{3.25c}$$

will maintain in the exit stream the same value as in the input stream. Furthermore, the variables Ω_j^* satisfy the consistency relationship:

$$\sum_{j>K+E} \Omega_j^* = 1 \tag{3.25d}$$

It is worth mentioning that the set of key species should be chosen so as to avoid the (fortuitous) situation $\Omega_{E0} = 1$, as it would result in undetermined Ω_{j0}^* values.

The definition (3.25c) for Ω_j^* can be used at any position and time. The qualifier "modified" is employed because the proportional constant $(1 - \Omega_E)^{-1}$ is a system variable and not a specified quantity, as in the definition of component variables in Sect. 3.3, and also because the set $\{\Omega_j^*\}$ excludes E component species. Recalling the definition of the Ω_j (Eq. 3.8), each ω_j (for $j > K+E$) is related to the associated Ω_j^* as (*cfr*. Equation 3.10a):

$$\omega_j = (1 - \Omega_E)\Omega_j^* + \sum_k \hat{\sigma}_{kj}\omega_k; \qquad j > K + E \tag{3.25e}$$

It should be noted that the global balances (3.24) and Eq. (3.25b) apply for any flow reactor with a single input-stream in SS, in particular for the systems studied in Sects. 3.6 and 3.7.

To explore the behaviour of the Ω_j^* replacing the Ω_j ($j > K + E$) along the tube, their governing equations should be first developed. Based on Eq. (3.22a), the following operations are made. First, for each $j > K+E$, $\Omega_j = (1-\Omega_E)\Omega_j^*$ is replaced. Second, the equations for all the significantly transferred species $K < j = e \leq K + E$ are added, and the result multiplied by Ω_j^*. Finally, this result is added to each of the equations previously obtained for $j > K + E$. It is thus obtained:

$$(1-\Omega_E)\left(\rho\frac{\partial\Omega_j^*}{\partial t}+G\frac{\partial\Omega_j^*}{\partial z}\right)= \\ ADTs-a_w\left[\mathbb{G}_{jw}-(G_w-\mathbb{G}_{Ew})\Omega_j^*\right];\qquad j>K+E \tag{3.26a}$$

where $\mathbb{G}_{Ew}=\sum_e\mathbb{G}_{ew}$ and $ADTs$ stands for "axial dispersion terms". Operating similarly on Eq. (3.22b), the following initial and boundary conditions are obtained:

$$\Omega_j^*(0,z)=\Omega_{jI}^*(z); \\ (1-\Omega_{E0})G_0\left(\Omega_j^*-\Omega_{j0}^*\right)_{z=0}=(ADTs)_{z=0};\qquad j>K+E \\ \left(\frac{\partial\Omega_j^*}{\partial z}\right)_{z=L}=0; \tag{3.26b}$$

The significant fact in Eq. (3.26a) is that in the transfer term $G_w\approx\mathbb{G}_{Ew}$ and $\mathbb{G}_{jw}\approx 0$ $(j>K+E)$ and then $\left[\mathbb{G}_{jw}-(G_w-\mathbb{G}_{Ew})\Omega_j^*\right]\approx 0$. The impact of the interfacial-transfer term on Ω_j^* is thus minimised. Therefore, as discussed in the following sub-section, situations like in Sects. 3.4.2.1 or 3.4.2.3 with a beneficial use of the Ω_j^*, can be identified. Instead, the balance equations for the species A_e, either Eqs. (3.21a, b) in terms of ω_e, or alternatively Eqs. (3.22a, b) in terms of Ω_e, should be solved simultaneously with Eq. (3.20a, b) for the key species.

Use of Variables Ω_j^* Under Steady State

In a transient state, the first two terms in Eq. (3.26a) will be non-zero. Hence, the Ω_j^* become subject to variations of G (Eq. 3.18a, $\partial\rho/\partial t+\partial G/\partial z+a_wG_w=0$), which can be highly significant, because of the dependence on the fluxes G_w and, hence on the mixture composition. The Ω_j^* will inherit this effect, and therefore, their use will not bring any evident benefit. Although reactions with diluted reactants can suppress such an effect, the discussion here is focussed on steady-state operations, when variations of G will not introduce serious limitations on the behaviour of the Ω_j^*.

First, if a common value D for the dispersion coefficients can be adopted, the term $ADTs$ in Eq. (3.26a) results:

$$ADTs=(1-\Omega_E)\frac{\partial}{\partial z}\left(\rho D\frac{\partial\Omega_j^*}{\partial z}\right)-2\rho D\frac{\partial\Omega_E}{\partial z}\frac{\partial\Omega_j^*}{\partial z} \tag{3.26c}$$

For $(ADTs)_{z=0}$ in Eq. (3.26b):

$$(ADTs)_{z=0}=\left[(1-\Omega_E)\frac{\partial}{\partial z}\left(\rho D\frac{\partial\Omega_j^*}{\partial z}\right)\right]_{z=0} \tag{3.26d}$$

In the SS ($\partial\Omega_j^*/\partial t=0$) and assuming that $G_{jw}=0$ is strictly valid for the species other than the A_e, the mass transfer terms in Eqs. (3.26a, b) will vanish and their solution, along with Eqs. (3.26c, d), are satisfied with $\Omega_j^* = \Omega_{j0}^*$. In this case, it can be said that variables Ω_j^* behave as *reaction/interface-transfer invariants.*

As a marginal note, it can be verified that for the ideal plug-flow assumption ($D_j = 0$), the component variables for the species flux rates, $\mathbb{G}_j = G_j - \sum_k \hat{\sigma}_{kj} G_k$, will behave as reaction invariants for $j > K+E$ and also avoid the indeterminacy of the Ω_j^* if eventually $\Omega_{E0} = 1$.

For a general SS with different values of D_j and/or low values $G_{jw} \neq 0$ for species A_j ($j > K + E$), the Ω_j^* will no longer behave as reaction invariants. However, under the frequent situation of a moderate effect of the $ADTs$ in Eqs. (3.26a, b), the deviation from the invariance condition can be small enough for trying the resolution procedure described in Sect. 3.4.2.2. To this end, some specific aspects should be considered:

i. The subset of the key-species equations, the total mass equation and the energy and momentum equations (when variations of T and P should be evaluated) is expanded with the equations for the component species A_e, either in terms of ω_e (Eqs. 3.21a, b) or in terms of Ω_e (Eqs. 3.22a, b). A module for the evaluation of the fluxes G_{jw} should be included. In a new iteration step, updated (fixed) values of Ω_j^* are used for the evaluation of ω_j ($j > K+E$) from Eq. (3.25e) (with Ω_e and ω_k as variables). The outcome from the step will be a set of updated values (ω_k, ω_e, G, T, P and G_{jw}) in each node. Updated values of ρ, D_j, Ω_E (and eventually μ), needed in a new iteration step for the Ω_j^*, will be also available.
ii. To update the values Ω_j^* it is not necessary to use the governing Eqs. (3.26a, b), because expressing the terms $ADTs$ is tiresome. Instead, Eqs. (3.22a, b) for $j > K+E$ (without the temporal term) can be used to update first the Ω_j in each node. These values should be corrected to maintain consistency, as discussed in the Appendix 2. Afterwards, values $\Omega_j^* = \Omega_j/(1 - \Omega_E)$ are calculated in each node (using the values Ω_E available from item (i), which will be used for a new iteration of the equation subset described in item (i).

Example 1: Axial dispersion model in steady state
The following set of gas-phase reactions and kinetics is considered:

$$\mathcal{R}_1: \mathrm{A} + \mathrm{B} \Leftrightarrow \mathrm{P}, \qquad r_1 = k_1\left[C_A C_B - \frac{C_P}{K_{C1}}\right]$$

$$\mathcal{R}_2: \mathrm{A} + \mathrm{P} \Leftrightarrow \mathrm{S} + \mathrm{W}, \quad r_2 = k_2\left[C_A C_P - \frac{C_S C_W}{K_{C2}}\right]$$

$$\mathcal{R}_3: 2\mathrm{A} + \mathrm{B} \Leftrightarrow \mathrm{S} + \mathrm{W}, \; r_3 = k_3\left[C_A^2 C_B - \frac{C_S C_W}{K_{C3}}\right]$$

As $\mathcal{R}_3 = \mathcal{R}_1 + \mathcal{R}_2$, the number of independent reactions is $K = 2$.

The reaction set is adapted from [12], where the catalytic hydrogenation ($A \equiv H_2$) of furfural (B) to furfuryl alcohol (P) is studied. Furfuryl alcohol can be further hydrogenated

Table 3.1 Molar masses (kg $kmol^{-1}$) and input composition

Species	A	B	P	S	W	I
m_j	2	96.1	98.1	82.1	18	100.2
y_{j0}	0.2	0.2	0	0	0	0.6

Table 3.2 Values of kinetic coefficients and equilibrium constants

k_1 (m^3 $kmol^{-1}$ s^{-1})	k_2 (m^3 $kmol^{-1}$ s^{-1})	k_3 (m^6 $kmol^{-2}$ s^{-1})	K_{C1} (m^3 $kmol^{-1}$)	K_{C2}	K_{C3} (m^3 $kmol^{-1}$)
1.816×10^2	3.708×10^1	1.073	4.628×10^2	1.328×10^3	6.146×10^5

to yield 2-methylfuran (S) and water (W). Reaction $\mathcal{R}_3$ accounts for a single step production of S and W from furfural. The mixture is diluted by the presence of chemically inert cyclopentyl methyl ether (I). Molar masses and mole fractions (assigned values) in the reactor input are shown in Table 3.1.

Species A and B are chosen as key species and, hence, P, S, and W are the component species. The canonical reactions for A and B and their net production rates are:

$$\mathcal{R}_A(= -\mathcal{R}_2): \qquad S + W \Leftrightarrow A + P, \quad r_A = -(r_1 + r_2 + 2r_3)$$
$$\mathcal{R}_B(= -2\mathcal{R}_1 + \mathcal{R}_3): 2P \Leftrightarrow B + S + W, \quad r_B = -(r_1 + r_3)$$

The molar stoichiometric coefficients of the component species in the canonical reactions are: $\sigma_{AP} = 1$, $\sigma_{BP} = -2$; $\sigma_{AS} = -1$, $\sigma_{BS} = 1$; $\sigma_{AW} = -1$, $\sigma_{BW} = 1$. It is recalled that in the conservation equations, $\hat{r}_j = m_j r_j$; $\hat{\sigma}_{kj} = (m_j/m_k)\sigma_{kj}$ are employed.

The reactions are processed in a homogeneous tubular reactor of length $L = 0.5$ m and inner diameter $d_t = 0.0762$ m, under essentially isothermal and isochoric conditions at $T = 190$ °C and $P = 2$ bar. The mass velocity is taken as $G = 0.3889$ kg m^{-2} s^{-1}.

The kinetic expressions, following the law of mass action, are hypothetical. Ideal gas behaviour is assumed and hence the equilibrium constant $K_{C,i}$ remain constant at the assumed uniform values of P and T. Values of kinetic coefficients and equilibrium constants displayed in Table 3.2 are chosen for the illustrative purposes of the example. Molar concentrations in the reaction rates are expressed as $C_j = \rho\omega_j/m_j$.

The mixture viscosity is assumed constant and estimated as $\mu = 1.482 \times 10^{-4}$ kg m^{-1} s^{-1}. Then, laminar flow arises: $Re = Gd_t/\mu = 200$.

The density at the reactor entry by using the ideal gas law is $\rho_0 = 4.233$ kg m^{-3}, and variations due to the molar change of the reactions (the reduction from reaction $\mathcal{R}_1$ is dominating) are evaluated as $\rho = \rho_0 \overline{m}/\overline{m}_0$, with the mean molar mass given by:

$$\overline{m} = \sum_j y_j m_j = \left(\sum_j \omega_j/m_j\right)^{-1}$$

The steady-state simulation will be made for three different situations concerning dispersion effects. As a reference, the first situation simply ignores dispersion and hence the plug-flow behaviour is assumed. The second one deals with the ADM but using a common dispersion coefficient D for all species. Finally, individual values D_j are considered in the third situation. In every situation the mass fraction of the inert species I remains constant, $\omega_I = \Omega_I = \omega_{I0}$. Hence, it is not further discussed.

Situation 1: Plug-Flow Model

The balance Eqs. (3.5a, b) are used for the key species, together with the relationships (3.8) and $\Omega_j = \Omega_{j0}$ (Eq. 3.9) for $j = \text{P, S, W}$. The equation set was solved by employing the routine *ode15s* of MATLAB®.

Situation 2: ADM with a Common Dispersion Coefficient D for All Species

Since the fluid-dynamic regime is laminar, Eqs. (3.15a, b) are used to evaluate the dispersion coefficients. In the present situation, it is taken the same molecular diffusivity in the mixture ($\mathfrak{D}$) for all species, which is chosen as the binary coefficient between species B and the inert species I, estimated as $\mathfrak{D}_{B,I} = 0.4375 \text{ cm}^2 \text{ s}^{-1}$. Thus, the common dispersion coefficient from Eqs. (3.15a, b) results at input conditions $D = 58.77 \text{ cm}^2 \text{ s}^{-1}$. The validity of Eq. (3.15b) to estimate the Taylor contribution is restrained to values $Pe_L = GL/(\rho D)$ higher than about 10. In the present case, $Pe_L \approx 7.8$, which is still acceptable for the purposes of this example. Variations of $u = G/\rho$ in the Taylor contribution along the tube, Eq. (3.15b), were considered.

The use of a common value D allows variables Ω_j to be reaction invariants (*Particular case 1*, Sect. 3.4.2.1), i.e., $\Omega_j = \Omega_{j0}$ for $j = \text{P, S, W}$.

For the key species A and B, Eqs. (3.20a, b) with $\partial\omega_k/\partial t = 0$ and without the mass transfer terms should be solved:

$$G\frac{d\omega_k}{dz} - \frac{d}{dz}\left(\rho D_k \frac{d\omega_k}{dz}\right) = \hat{r}_k; \qquad k = \text{A, B} \tag{E3.1a}$$

$$\begin{aligned} &G_0(\omega_k - \omega_{k0})_{z=0} - \left(\rho D_k \frac{d\omega_k}{dz}\right)_{z=0} = 0; \\ &\left(\frac{d\omega_k}{dz}\right)_{z=L} = 0; \end{aligned} \tag{E3.1b}$$

where $D_A = D_B = D$.

For this situation (and also for Situation 3 below), a grid of 150 nodes uniformly spaced was employed to discretise Eqs. (E3.1a, b). The derivatives were approximated by using a scheme of order 3–4 (five nodal values involved in each approximation). The set of algebraic equations was solved using the MATLAB® routine *fsolve*.

Situation 3: MDA with Individual Values D_j for Each Species

Equation (3.16) is used to estimate individual values $\mathfrak{D}_j$ from values of binary molecular coefficients $\mathfrak{D}_{j,i}$. These are evaluated by considering the dependence $\mathfrak{D}_{j,i} \propto (1/m_{j,i})^{1/2}$,

Table 3.3 Values of $\mathfrak{D}_j$ and D_j [m^2 s^{-1}] in the input stream

	A	B	P	S	W
$10^4\ \mathfrak{D}_j$	1.840	0.4074	0.4079	0.4150	0.5684
$10^4\ D_j$	15.71	63.05	62.97	61.91	45.47

where $1/m_{j,i} = 1/m_j + 1/m_i$, and employing the reference value $\mathfrak{D}_{\mathrm{B,I}} = 0.4375\ \mathrm{cm}^2\ \mathrm{s}^{-1}$ used in Situation 1:

$$\mathfrak{D}_{j,i} = [\mathfrak{D}_{\mathrm{B,I}}/(1/m_{\mathrm{B,I}})^{1/2}](1/m_{j,i})^{1/2}$$

Then, Eq. (3.16) is employed to evaluate $\mathfrak{D}_j$ and, finally, each specific value D_j follows from Eq. (3.15a, b). Values at the tube entry are displayed in Table 3.3, where it can be appreciated that the Taylor contribution is dominant. Also, a large difference between D_{A} (for H_2) and any of the remainder values arises.

Because of limitation (A) in Sect. 3.2, variables Ω_j no longer behave as reaction invariants. Two alternatives have been used to find the solution for this situation.

In the first place, the conservation equations of all reacting species (i.e., Eqs. 3.20a, b and Eqs. 3.21a, b without the transient and mass transfer terms) were simultaneously solved.

The second alternative was the use of the resolution strategy described in Sect. 3.4.2.2. In this case, the equation subset concerning the key species are just Eqs. (E3.1a, b), while the subset for the component variables Ω_j (j = P, S, W) follows from Eqs. (3.22a, b) without the transient and mass transfer terms:

$$G\frac{d\Omega_j}{dz} - \frac{d}{dz}\left(\rho D_j \frac{d\Omega_j}{dz}\right) = \frac{d}{dz}\left[\rho \sum_k \hat{\sigma}_{kj}(D_j - D_k)\frac{d\omega_k}{dz}\right]; \quad j = \mathrm{P, S, W} \qquad \text{(E3.2a)}$$

$$G_0(\Omega_j - \Omega_{j0})_{z=0} - \left(\rho D_j \frac{d\Omega_j}{dz}\right)_{z=0} = \left[\rho \sum_k \hat{\sigma}_{kj}(D_j - D_k)\frac{d\omega_k}{dz}\right]_{z=0}; \qquad \text{(E3.2b)}$$
$$\left(\frac{d\Omega_j}{dz}\right)_{z=L} = 0;$$

The relatively high values of axial Peclet numbers $Pe_{L,j} = GL/(\rho D_j)$ (higher than 7) suggest that dispersion effects will be moderate enough to accomplish the conditions of Sect. 3.4.2.3 (particular case 2) and, consequently, the resolution strategy of Sect. 3.4.2.2 can be expected to be efficient.

Results

For the presentation of the results, it is convenient to employ the variables $\psi_j = \omega_j/m_j$ (moles of A_j per unit mass of mixture). In particular, furfural (B) conversion, selectivity and yield of furfuryl alcohol (P), x_{B}, Φ_{BP} and η_{BP} respectively, at the reactor exit are expressed as (with $\psi_{\mathrm{P}0} = 0$):

Table 3.4 Conversion of B, selectivity, and yield of P from B

Model	x_B	Φ_{BP}	η_{BP}
Chemical equilibrium	0.5069	0.0431	0.0219
Plug flow	0.5941	0.5441	0.3232
MDA (*D common for all species*)	0.6013	0.5930	0.3566
MDA (D_j *specific for each species*)	0.5976	0.5548	0.3316

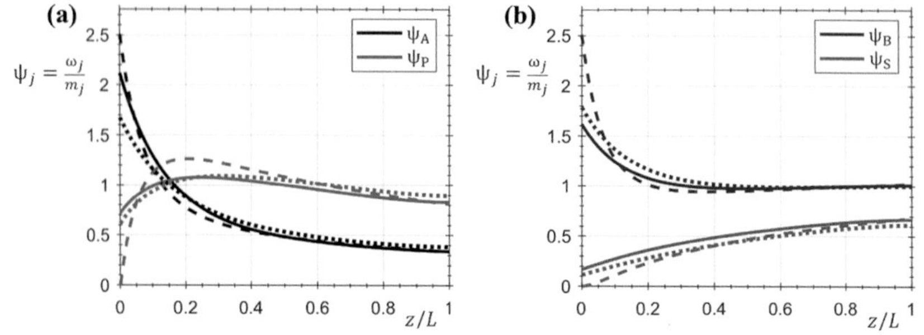

Fig. 3.1 Concentration profiles, ψ_j [mol kg^{-1}]. Panel **a** for A and P. Panel **b** for B and S. Plug-flow model (dashed curves), ADM with common D (dotted curves), ADM with individual D_j (continuous curves). The profile ψ_W (not shown) is quite similar to ψ_S

$$x_B = 1 - \frac{\psi_B}{\psi_{B0}}; \quad \Phi_{BP} = \frac{\psi_P}{\psi_{B0} - \psi_B}; \quad \eta_{BP} = \frac{\psi_P}{\psi_{B0}} = \Phi_{BP} x_B$$

Table 3.4 shows the results from the three situations considered and also values at chemical equilibrium (i.e., when $L \to \infty$).

The low selectivity and yield, Φ_{BP} and η_{BP}, at chemical equilibrium are significantly in contrast to the values at the finite length $L = 0.5$ m. The low equilibrium values are due to the fact that reaction $\mathfrak{R}_1$ is thermodynamically disfavoured with respect to reaction $\mathfrak{R}_2$ or $\mathfrak{R}_3$. However, the opposite happens kinetically. Therefore, Φ_{BP} and η_{BP} present a spatial maximum, as can be appreciated from the profile of ψ_P given in Fig. 3.1a. Another consequence of the trade-off between thermodynamic and kinetic effects is that the reaction $\mathfrak{R}_1$ reverses (r_1 changes sign) and, hence, the profile of ψ_B shows a minimum (a maximum of x_B), which is present in the results of the three models, Fig. 3.1b, although barely appreciable at the scale used therein.

Figure 3.1 shows that, in both situations "2" and "3", the effect of dispersion is significant close to the reactor entry whereas clearly weakens towards the exit. This behaviour is reflected by comparing the results in Table 3.4 with those of the plug-flow model, although the values of Φ_{BP} and η_{BP} for situation 2 (common value of D) show a considerable difference of about 10%. It can also be appreciated in Table 3.4 that the three

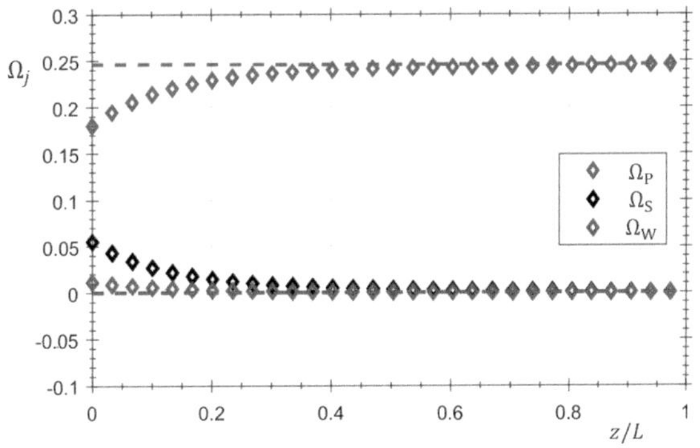

Fig. 3.2 Profiles of the Ω_j from the ADM with individual D_j (diamonds) and values Ω_{j0} (dashed lines). Note that $\Omega_{S0} = \Omega_{W0}$

indicators of process performance x_B, Φ_{BP} and η_{BP} increase with dispersion, a fact that seems to contradict expectation. A possible explanation is that the plug-flow model predicts a faster increase in the yield of product P (see profile of ψ_P in Fig. 3.1a) and hence an early appearance of the maximum occurs. Then, a longer length is available for thermodynamics to come into effect.

However, it can be seen in Fig. 3.1a that the maximum of ψ_P for situation 3 (individual values of D_j) takes place closer to the position from the plug-flow model. Thus, the results of both situations in Table 3.4 only differ slightly. This feature is due to the very low H_2 dispersion coefficient ($D_A = 15.71$ cm^2/s, Table 3.2), allowing values of r_1 similar to those of the plug-flow model. Instead, the common value $D = 58.77$ cm^2/s in situation 2 is much higher and thus noticeably reduces the rate r_1.

In the example, the density changes affect the dispersion coefficients and the reaction rates. However, the variations are not large, due to the presence of the inert, with a maximum value $\rho/\rho_0 \approx 1.15$, close to the reactor exit for the three situations studied. Therefore, the discussed results are barely influenced by density effects.

Profiles of variables Ω_j for Situation 3 are shown in Fig. 3.2. Their departure from the values Ω_{j0} (reaction invariants for Situations 1 and 2) are only significant in the region close to the reactor entry, promoted by the differences $(D_j - D_k)$ in Eqs. (E3.2a, b), as discussed in Sect. 3.4.1. The low value of D_A (Table 3.3) for the key species A introduces largest differences on $(D_j - D_k)$: $D_j/D_A \approx 4$ for the component species j = P, S and $D_W/D_A \approx 2.9$. Consequently, Ω_W shows the smallest variations (Fig. 3.2).

Additionally, species A undergoes the fastest concentration changes close to the entry and hence the term $[(D_j - D_A)d\omega_A/dz]$ in Eqs. (E3.2a, b) is significantly large. In spite of this feature, the use of the resolution strategy described in Sect. 3.4.2.2 behaved correctly. The computing time was 35% lower than for the simultaneous solution of all species balances, while the number of total iterations (three) was not modified. The saving of computing time is significant, notwithstanding the low number of species involved.

3.5 Reaction Systems with Diffusive Transport

The ADM is not conceived to describe the behaviour of the variables close to the solid boundaries, as only changes in the axial direction of the flow are considered. Instead, the models in Sects. 3.6 and 3.7 are intended to evaluate the distribution of the state variables in the whole reactor domain. Thus, the presence of the solid surfaces, e.g., the vessel walls, should be explicitly considered, since they strongly impair convection and therefore the species fluxes—of special importance in the normal direction to the surfaces—are dominated by molecular diffusion. In heterogeneous catalytic reactors (Sect. 3.7), such diffusion-dominated fluxes determine the capacity of interfacial transfer on the fluid stream side. Moreover, inside the pores of the catalyst, the fluxes are strictly of diffusion nature.

Examples of simplified diffusion/reaction systems are undertaken in this section with the main objective of drawing conclusions about the significance of the variables Ω_j.

Case I: Reaction inside a stagnant slab

In first place, it is considered a catalytic porous slab with one of its surfaces contacting an external fluid. The reactions are catalysed by the active walls of the pores, with net production rates $\hat{r}_j$ expressed per unit volume of the slab, and uniform slab porosity ε_p. This example exhibits the basic behaviour of a catalyst particle or, even with a closer geometrical resemblance, of the catalytic *washcoat* in monolithic reactors (Sect. 3.9).

The diffusion mass-fluxes j_j normal to the slab cross-section are expressed by:

$$j_j = -\rho \mathfrak{D}_{j,ef} \partial \omega_j / \partial \zeta, \tag{3.27}$$

where ζ is the normal coordinate and $\mathfrak{D}_{j,ef}$ are effective diffusivities. Both properties ρ and $\mathfrak{D}_{j,ef}$ are assumed to remain constant.

It is noted that in problems with concentrated reactive species, it will be necessary to employ more accurate (and complex) expressions for the fluxes j_j, e.g., based on Maxwell–Stefan equations (see e.g., Kaza and Jackson [13] for a discussion on solution procedures). Nonetheless, the use of Eq. (3.27) will suffice for the present purpose.

The case of steady-state conditions will be considered here, while Appendix 3 treats the more general situation of transient behaviour. The relevant conclusions from the Appendix 3 will be briefly summarised later on.

Conservation equations in SS for the species and boundary conditions are:

$$\rho \mathfrak{D}_{j,ef} \frac{d^2 \omega_j}{d\zeta^2} = -\hat{r}_j = -\sum_k \hat{\sigma}_{kj} \hat{r}_k; \qquad j > K \tag{3.28a}$$

$$\begin{aligned} (d\omega_j/d\zeta)_{\zeta=\ell} &= 0; \\ \omega_j(\zeta = 0) &= \omega_{j0}; \end{aligned} \tag{3.28b}$$

$$\rho\mathfrak{D}_{k,ef}\frac{d^2\omega_k}{d\zeta^2} = -\hat{r}_k; \qquad k = 1, \ldots, K \tag{3.29a}$$

$$\begin{aligned}(d\omega_k/d\zeta)_{\zeta=\ell} &= 0;\\ \omega_k(\zeta = 0) &= \omega_{k0};\end{aligned} \tag{3.29b}$$

where ω_{j0} $(1 \leq j \leq S)$ are the values at the slab surface in contact with the external fluid.

By making the operation between equations $[(3.28a)_{j>K} - \sum_k (3.29a)_k\hat{\sigma}_{kj}]$, it is obtained for each $j > K$:

$$\frac{d^2\left(\mathfrak{D}_{j,ef}\omega_j - \sum_k \hat{\sigma}_{kj}\mathfrak{D}_{k,ef}\omega_k\right)}{d\zeta^2} = 0; \qquad j > K \tag{3.30}$$

Upon integrating Eq. (3.30) twice and accounting for both boundary conditions, it is obtained:

$$\mathfrak{D}_{j,ef}\omega_j - \sum_k \hat{\sigma}_{kj}\mathfrak{D}_{k,ef}\omega_k = \mathfrak{D}_{j,ef}\omega_{j0} - \sum_k \hat{\sigma}_{kj}\mathfrak{D}_{k,ef}\omega_{k0}; \qquad j > K \tag{3.31}$$

The reaction invariants arisen from Eq. (3.31) are:

$$\mathbb{J}_j = \mathfrak{D}_{j,ef}\omega_j - \sum_k \hat{\sigma}_{kj}\mathfrak{D}_{k,ef}\omega_k; \qquad j > K \tag{3.32}$$

which remain constant and hence can be evaluated at the external slab surface:

$$\mathbb{J}_j = \mathbb{J}_{j0}; \qquad j > K \tag{3.33}$$

In general, the $\mathbb{J}_j$ can be defined as *diffusive component variables*, as they are clearly related to the diffusion capability of the species. Given the value $\mathbb{J}_{j0}$, the corresponding mass fraction ω_j at any internal position can be evaluated:

$$\mathfrak{D}_{j,ef}\omega_j = \mathbb{J}_{j0} + \sum_k \hat{\sigma}_{kj}\mathfrak{D}_{k,ef}\omega_k; \qquad j > K \tag{3.34}$$

Rearranging expressions (3.31) in terms of the component variables Ω_j:

$$\Omega_j - \Omega_{j0} = \sum_k \hat{\sigma}_{kj}\left(\frac{\mathfrak{D}_{k,ef}}{\mathfrak{D}_{j,ef}} - 1\right)(\omega_k - \omega_{k0}); \qquad j > K \tag{3.35}$$

It follows from Eq. (3.35) that the values Ω_j inside the slab will depart from Ω_{j0} in proportions to $[(\mathfrak{D}_{k,ef}/\mathfrak{D}_{j,ef}) - 1]$ and, hence, they will depend on the progress of the reactions by virtue of the factors $(\omega_k - \omega_{k0})$. As significant differences between effective diffusivities will be frequently found, it is concluded that, in general, the variables Ω_j will not be reaction invariants, as they will be subject to a significant effect of the reaction rates.

Although in the previous paragraphs the values ω_{j0} at the slab surface, and hence the $\mathbb{J}_{j0}$, have been assumed known, mass transfer resistances in the external fluid cannot be frequently ignored. Such resistances can be evaluated by resorting to correlations of mass transfer coefficients, which allow relating the known values ω_{jF} in the bulk of the fluid phase with those on the surface slab, ω_{j0}. The values $\mathbb{J}_{j0}$ can thus be evaluated, but as shown in Appendix 3, they will depend on the reaction rates in the slab. As a consequence, the variables $\mathbb{J}_j$ (see Eq. 3.33) will no longer behave as reaction invariants. Nonetheless, as the $\mathbb{J}_j$ remain uniform in the slab, their use is beneficial for the resolution of the balances in steady state, as discussed in Appendix 3.

Instead, it is shown in Appendix 3 that in a transient state, neither the $\mathbb{J}_j$ nor the Ω_j will behave as reaction invariants, irrespective of external mass resistances.

For a layer of stagnant fluid with homogeneous reactions, the formulation presented above is applicable with diffusivities $\mathfrak{D}_j$ replacing $\mathfrak{D}_{j,ef}$ and $\hat{r}_j$ standing for the rates of the homogeneous reactions. The conclusions, particularly those concerning the variables Ω_j, are equally valid. This view will be relevant in the next section.

Case II: Diffusion through a stagnant film and reaction at the interface

The *film model* can be used as an elementary description of the interfacial mass transfer from a fluid stream to the surface of a catalytic body, as a catalytic pellet or the washcoat in a monolithic reactor. The species are assumed to diffuse from the bulk of the fluid to the interface through a stagnant film of thickness ℓ. The formulation in Eqs. (3.28a–3.35) can be conveniently used in this situation by assimilating the coefficients $\mathfrak{D}_{j,ef}$ to $\mathfrak{D}_j$, suppressing the reactions within the film and considering as boundary conditions known values ω_{j0} at $\zeta = 0$ (fluid bulk) and $\mathfrak{D}_j(\partial\omega_j/\partial\zeta)_{\zeta=\ell} = \hat{r}_{j,sup}$ at $\zeta = \ell$ (interface), where $\hat{r}_{j,sup}$ is the net production rate of A_j per unit surface area.

Equation (3.28a) with the r.h.s. equal to zero applies. Equation (3.33) will still hold and at both ends of the film it is verified that $\mathbb{J}_{j,\ell} = \mathbb{J}_{j0}$. Therefore, the behaviour of the variables $\mathbb{J}_j$ is maintained. Regarding the variables Ω_j, the equality $\mathbb{J}_{j,\ell} = \mathbb{J}_{j0}$ allows writing:

$$\mathfrak{D}_j(\Omega_{j,\ell} - \Omega_{j0}) = -\ell \sum_k \hat{\sigma}_{kj}\left[\left(\frac{\mathfrak{D}_j}{\mathfrak{D}_k}\right) - 1\right]\hat{r}_{k,sup}; \qquad j > K \tag{3.36}$$

For given values Ω_{j0}, the relevant values $\Omega_{j,\ell}$ at the interface will in general depend on $\hat{r}_{k,sup}$ and therefore can significantly depart from a reaction-invariance behaviour.

It is reminded that the film model cannot predict the actual rate of the mass transfer, $\rho\mathfrak{D}_j(\omega_{j0} - \omega_{j,\ell})/\ell$, as the thickness ℓ depends on the fluid-dynamics around the surface, including the effect of the Schmidt numbers $Sc_j = \mu/(\rho\mathfrak{D}_j)$. However, the previous discussion reveals that if certain models (see Sect. 3.9) are able to predict mass transfer rates (i.e., values of ℓ), the use of the variables Ω_j are not likely to introduce computational benefits.

Summing up, for systems in which diffusion transport is dominant, the main conclusions are:

- the use of variables Ω_j will not bring in general computational benefits, neither in the SS nor in a transient state, because of their dependence on the state variables, caused by differences in diffusion coefficients.
- Variables $\mathbb{J}_j$ can be efficiently employed in SS, but not in a transient state.

3.6 Homogeneous Reactors with Turbulent Flow in More Than One Spatial Coordinate

Homogeneous reaction systems with properties changing in more than one spatial coordinate are considered here. With respect to the ADM, transversal profiles of the variables should be accounted for when average values cannot accurately describe the reactor behaviour, for instance, when heat is exchanged through the reactor walls. A more complex reactor geometry than that of a tube with constant cross-section area will also require a 3D description of the system.

Although the conservation equations presented in this section are suitable for laminar flow, the discussion concerning the behaviour of variables Ω_j are restrained to operations in turbulent regime, because of the reasons explained in Sect. 3.6.1.

It should be noted that the meaning of the variables in the conservation equations for turbulent flow correspond to time-smoothed quantities and transport properties are strongly augmented by turbulent mixing [5]. In particular, the dispersion mass-flux vector of a species A_j, $\underline{j}_j$, is expressed as:

$$\underline{j}_j = -\rho D_j \nabla \omega_j \tag{3.37}$$

where D_j is an effective dispersion coefficient:

$$D_j = \mathfrak{D}_j + D^t \tag{3.38}$$

In Eq. (3.38), $\mathfrak{D}_j$ is the molecular diffusivity of the species A_j in the mixture, which can be approximated as in Eq. (3.16) from values of binary diffusivities, and D^t is the turbulent contribution that will be considered here as a common value for all species. Although it is acknowledged that D^t depends to some degree on the individual Schmidt numbers, $Sc_j = \mu/(\rho\mathfrak{D}_j)$, it appears that at present there is no plausible relationship to establish such a dependence (see e.g., Gualtieri et al. [14]).

While D^t in Eq. (3.38) is the major contribution to dispersion far enough from the walls, it strongly decreases in regions close to them. Because of the non-slip condition on

Fig. 3.3 Radial profiles of D (curve identification explained in the text). Original drawing presented as Fig. 24 in Feng et al. [15]; reproduced with permission of the copyright owner, Elsevier Ltd.

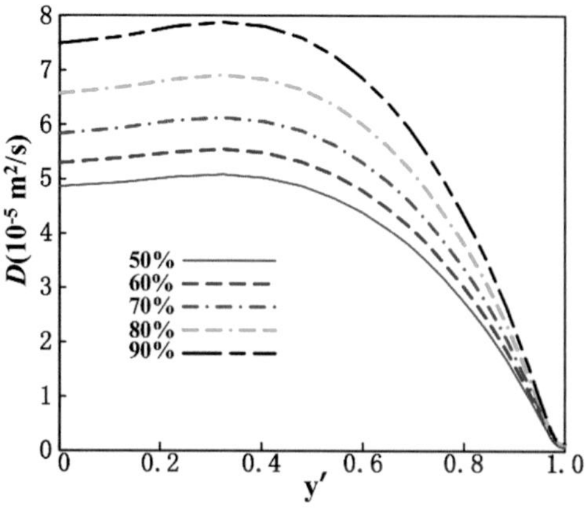

the walls (zero tangential velocity), a thin *viscous sublayer* can be identified, where $D^t = 0$ and hence $D_j = \mathfrak{D}_j$ in Eq. (3.38). Outside the viscous sublayer, D^t starts increasing up to some point where $D^t \gg \mathfrak{D}_j$ is reached. This region can be described as *transition layer*. Further apart from the wall, a third layer can be distinguished (*turbulent layer*), before finally reaching the *turbulent nucleus*, where the properties become independent of wall effects (such a description also applies to other transport properties, such as effective viscosity and thermal conductivity, see e.g., Bird et al. [5]).

An example of radial profiles of D_j is given in Fig. 3.3, from Feng et al. [15] study of the pyrolysis of a hydrocarbon fuel at supercritical conditions. The data in Fig. 3.3 were calculated with a value $Re = 4.4 \times 10^4$ at different levels of the reactant conversion (per cent values in the inset), corresponding to increasing temperature levels (approximately, 906, 933, 954, 971, 986 K) along the reactor axis. It can be appreciated that the molecular diffusivity only becomes significant (values of $\mathfrak{D}_j$ around 0.1×10^{-5} m^2/s) at very short distances from the wall ($y' = 1$ in Fig. 3.3).

To write conservation equations for the total mass and for each species, it is assumed that the stream enters the reactor across a section with uniform distributions of velocity and state variables. For the species balances, Danckwerts's boundary conditions are applied.

For total mass conservation,

$$\frac{\partial \rho}{\partial t} + \nabla \cdot \underline{G} = 0; \tag{3.39a}$$

$$G_n\left(t, \underline{z}_0\right) = G_0(t); \tag{3.39b}$$

where $\underline{G} = \rho \underline{v}$ is the vector of mass velocities, $\underline{z}_0$ is the vector of spatial coordinates (z_1, z_2, z_3) spanning the entry section, suffix "n" denotes a direction locally normal to a given surface and G_0 is a uniform and known value, which may be time dependent.

In the case of permeable walls,

$$[G_n]_w = G_w \tag{3.40}$$

where $G_w = \sum_j G_{jw}$ is the total mass flux through the wall and G_{jw} is the mass flux of species A_j. Similar comments as made in Sect. 3.4 apply for the fluxes G_{jw}.

The species balances are:

$$\frac{\partial}{\partial t}(\rho\omega_j) + \nabla\cdot\left(\omega_j \underline{G}\right) - \nabla\cdot\left(\rho D_j \nabla\omega_j\right) = \hat{r}_j \tag{3.41}$$

The rates $\hat{r}_j$ in Eq. (3.41) are also time-smoothed quantities. In the frequent case of non-linear kinetics, the average of instantaneous values should be evaluated, rather than using the average values of the state variables in the kinetic expressions. An example of such evaluations is presented in Appendix 4. In general, the evaluation of $\hat{r}_j$ will require a considerable extra effort, although in the end, the time-smoothed values of the state variables will be involved, along with a measure of the turbulence intensity. It is noted that the stoichiometric relationships $\hat{r}_j = \sum_k \hat{\sigma}_{kj}\hat{r}_k$ will remain valid.

By combining Eq. (3.41) and the total mass balance (3.39a), it is obtained:

$$\rho\frac{\partial\omega_j}{\partial t} + \underline{G}\cdot\nabla\omega_j - \nabla\cdot\left(\rho D_j \nabla\omega_j\right) = \hat{r}_j \tag{3.42a}$$

The initial condition is written as:

$$\omega_j(0, \underline{z}) = \omega_{jI}(\underline{z}) \tag{3.42b}$$

The Danckwerts' boundary conditions at the entry ($\underline{z}_0$) and exit ($\underline{z}_s$) sections are:

$$\begin{aligned} &\left[G_0(\omega_j - \omega_{j0}) - \rho D_j \nabla_n \omega_j\right]_{\underline{z}_0} = 0; \\ &\left(\nabla_n \omega_j\right)_{\underline{z}_s} = 0 \end{aligned} \tag{3.42c}$$

In the presence of permeable walls, $G_{jw} = [G_n\omega_j - \rho D_j \nabla_n \omega_j]_w$. Using Eq. (3.40):

$$-[\rho D_j \nabla_n \omega_j]_w = G_{jw} - \omega_j(\underline{z}_w) G_w \tag{3.42d}$$

By making the operation between equations $[(3.42\text{a-d})_{j>K} - \sum_k (3.42\text{a-d})_k \times \hat{\sigma}_{kj}]$, the governing equations for the component variables Ω_j (Eq. 3.8) are obtained. Then, the system (3.42a–d) can be split in a subset for the key species and another for Ω_j:

<u>For the key species</u>, $k = 1, \ldots, K$:

$$\rho\frac{\partial\omega_k}{\partial t} + \underline{G}\cdot\nabla\omega_k - \nabla\cdot(\rho D_k \nabla\omega_k) = \hat{r}_k \tag{3.43a}$$

$$\omega_k(0, \underline{z}) = \omega_{kI}(\underline{z}) \tag{3.43b}$$

$$[G_0(\omega_k - \omega_{k0}) - \rho D_k \nabla_n \omega_k]_{\underline{z}_0} = 0; \quad (\nabla_n \omega_k)_{\underline{z}_s} = 0 \tag{3.43c}$$

$$-[\rho D_k \nabla_n \omega_k]_w = G_{kw} - \omega_k(\underline{z}_w) G_w \tag{3.43d}$$

For the component variables, $j > K$:

$$\rho \frac{\partial \Omega_j}{\partial t} + \underline{G} \cdot \nabla \Omega_j - \nabla \cdot (\rho D_j \nabla \Omega_j) = \nabla \cdot [\rho \sum_k \hat{\sigma}_{kj} (D_j - D_k) \nabla \omega_k] \tag{3.44a}$$

$$\Omega_j(0, \underline{z}) = \Omega_{jI}(\underline{z}) \tag{3.44b}$$

$$\left[G_0(\Omega_j - \Omega_{j0}) - \rho D_j \nabla_n \Omega_j - \rho \sum_k \hat{\sigma}_{kj} (D_j - D_k) \nabla_n \omega_k \right]_{\underline{z}_0} = 0; \quad (\nabla_n \Omega_j)_{\underline{z}_s} = 0 \tag{3.44c}$$

$$-\left[\rho D_j \nabla_n \Omega_j + \rho \sum_k \hat{\sigma}_{kj} (D_j - D_k) \nabla_n \omega_k \right]_w = \mathbb{G}_{jw} - \Omega_j(\underline{z}_w) G_w \tag{3.44d}$$

In Sect. 3.4 for the ADM, the mass velocity G is only subject to axial variations due to temporal density changes or mass transfer through the walls (Eq. 3.18a). In the present systems, even in steady state and without mass transfer, the components of the mass velocity vector $\underline{G}$ will have to be evaluated by solving the conservation equations of total mass and momentum. As these equations involve fluid properties as density ρ and viscosity μ, the species and energy balances should be simultaneously included in the resolution. Thus, it should be taken into account that, in general, the distribution of the component of $\underline{G}$ in the system will depend on the state variables. An exception is the case of diluted enough reactants when constant values of ρ and μ can be assumed.

3.6.1 Limitations for the Existence of Reaction Invariants. Turbulent Flow in More Than One Spatial Coordinate

Equations (3.44a–d) present the same general limitations (A), (B) and (C) given in Sect. 3.4.1 for the behaviour of variables Ω_j as reaction invariants. Concerning limitation (C), the dependence of $\underline{G}$ on the state variables should be added to those of ρ and D_j.

As variations in more than one spatial coordinate are considered, Eqs. (3.43a–d) and (3.44a–d) can be employed for systems involving multiple input streams of different composition. Boundary conditions as in Eqs. (3.43c) and (3.44c) should be written for each entry section. If the streams are identified by a subscript α, there will be in particular different sets of entry values $\Omega_{j0,\alpha}$, which will lead necessarily to spatial variations of Ω_j in Eq. (3.44a), *i.e.*, $\nabla\Omega_j \neq 0$. This fact imposes the dependence of Ω_j on ρ, D_j and $\underline{G}$, hence on the state variables, and thus preventing them from behaving as reaction invariants, even in the SS. This conclusion will apply in general for any system with multiple stream inputs and can be added as limitation (D). On the contrary, whenever ρ, D_j and $\underline{G}$ can be assumed independent of the state variables, multiple input streams will not introduce any limitation on the behaviour of variables Ω_j. Given the fact that multiple input streams are not usually found in practice, they will not be further considered in this chapter.

The second particular feature of the present system concerns the regions close to the walls, where the convective transport of the species is hampered by the low values of the components of $\underline{G}$. Therefore, dispersion transport in the direction normal to the wall becomes significant to supply or remove the species involved in the chemical reactions. Simultaneously, as in the regions close to the wall the turbulent contribution D^t is weaken, as previously explained, molecular diffusion will be important. The situation resembles the problem discussed as *Case I* in Sect. 3.5, where it was concluded that variables Ω_j can depend considerably on the state variables (due to different values of $\mathfrak{D}_j$) and hence their use is not expected to provide concrete benefits in those regions. Moreover, in problems with permeable walls, the present model allows expressing the interface transfer rates, which can also depend strongly on the $\mathfrak{D}_j$ (as similarly concluded in *Case II* of Sect. 3.5).

The relevance of the discussed behaviour is set by the turbulence intensity. High Reynolds numbers will reduce the size of the region in which the transverse diffusion transport is significant, as can be the case illustrated in Fig. 3.3. On the contrary, if laminar flow were considered, the transversal transport governed by molecular diffusion would extend over the whole cross-section and therefore the practical significance of variables Ω_j would be strongly reduced. This is the reason why the treatment in this section has been focussed on turbulent flow.

3.6.2 Particular Cases in Systems with Turbulent Flow and More Than One Spatial Coordinate

Particular situations in which the use of Ω_j presents computational benefits are described next, based on the similar cases discussed in Sects. 3.4.2.1, 3.4.2.3 and 3.4.2.4.

3.6.2.1 Particular Case 1: Variables Ω_j Behave as Reaction Invariants

As for the ADM (Sect. 3.4.2.1), the suppression of limitations (A) and (B) are necessary conditions for the Ω_j to behave as reaction invariants. The assumption $D_j = D$ necessary to avoid limitation (A) is more stringent in the present system, because of the transverse transport effect discussed in Sect. 3.6.1, but will be sound at high enough Reynolds numbers, when the turbulent contribution D^t is dominant over essentially the whole cross-section, as discussed above. For numerical computations, a suitable value of the molecular diffusivity $\mathfrak{D}$ common for all species can be used in Eq. (3.38). Equation (3.44a–d) becomes:

$$\rho\frac{\partial\Omega_j}{\partial t}+\underline{G}\cdot\nabla\Omega_j-\nabla\cdot(\rho D\nabla\Omega_j)=0 \tag{3.45a}$$

$$\Omega_j(0,\underline{z})=\Omega_{jI}(\underline{z}) \tag{3.45b}$$

$$\begin{aligned}&\left[G_0\left(\Omega_j-\Omega_{j0}\right)-\rho D\nabla_n\Omega_j\right]_{\underline{z}_0}=0;\\&\left(\nabla_n\Omega_j\right)_{\underline{z}_s}=0\end{aligned} \tag{3.45c}$$

$$[\nabla_n\Omega_j]_w=0 \tag{3.45d}$$

In steady state, Eqs. (3.45a–d) is satisfied with $\Omega_j=\Omega_{j0}$. Then, Eqs. (3.43a–d) for the key species can be solved by employing for the component species $\omega_j=\Omega_{j0}+\sum_k\hat{\sigma}_{kj}\omega_k$.

For a transient state, the Ω_j will behave as reaction invariants if ρ, D and $\underline{G}$ can be assumed independent of the state variables, thus eliminating the limitation (C) in Sect. 3.4.1.

3.6.2.2 Resolution Procedure for Small Deviations of Variables Ω_j as Reaction Invariants

The resolution procedure proposed in Sect. 3.4.2.2 when the variables Ω_j do not strictly behave as reaction invariants, because of limitations (A) and (C) (Sect. 3.4.1), can be used for the present system, but subject to a modification explained next. As discussed in Sect. 3.4.2.2, such a resolution procedure requires for being efficient that variables Ω_j departs moderately from their behaviour as reaction invariants. This condition will not be fulfilled in regions close to the wall, as explained in Sect. 3.6.1, and therefore introduce the risk of misbehaviour in the resolution procedure.

A possible approach to avoid such a risk is to divide the domain in a "wall region", where the radial transport is significantly dependent on molecular diffusion, and a "turbulent region" in the rest of the domain. Continuity conditions on the boundary between the zones should be applied. In the wall zone, all the species balances are solved simultaneously and in the turbulent region the proposed procedure based on the alternate iterations of the Ω_j equations (i.e., Eqs. 3.44a–d) and of the remainder governing equations (including Eqs. 3.43a–d for the key species). This strategy will be more efficient the smaller the size of the wall region.

The boundary between both regions can be defined by estimating distances from the walls at which the differences between the D_j, as introduced by different molecular diffusivities (Eq. 3.38), become less than a certain level (e.g., 10–20%). For the example in Fig. 3.3, this criterion will lead to a turbulent region spanning from the axis up to around 95% of the tube radius.

3.6.2.3 Particular Case 2: Variables Ω_j Closely Behave as Reaction Invariants

The resolution procedure just described above can be applied in steady state provided that the wall is impermeable ($G_{jw} = 0$), to avoid limitation (B), as in the case of the ADM. Since the use of variables Ω_j is restricted to the turbulent region, differences in D_j, if any, will not be of concern. Instead, for the safe application in a transient state, temporal variations of ρ, D_j and $\underline{G}$ should be moderate, particularly in the turbulent region.

3.6.2.4 Particular Case 3: Selective Interfacial Mass Transfer in Steady State

As in Sect. 3.4.2.4 for the ADM, this case significantly restrains the mass transfer rates through the walls to a number E of component species A_e ($K < e \leq K + E$). The total mass flux of these species will satisfy $G_{Ew} \approx G_w$, where $G_{Ew} = \sum_e G_{ew}$.

Basically, the discussion in Sect. 3.4.2.4 for the ADM in terms of the *modified component variables* $\Omega_j^* = \Omega_j/(1 - \Omega_E)$ (Eq. 3.25c, $\Omega_E = \sum_e \Omega_e$), for the component species A_j ($j > K + E$) that are only marginally transferred through the interface, remain valid. Next, the main conclusions are revisited, along with some specific considerations for the present system.

In general, the usefulness of Ω_j^* is restricted to the steady state. In Sect. 3.4.2.4, it was shown that if in addition a common dispersion coefficient D can be applied and if the interfacial transfer of species A_j ($j > K + E$) can be strictly ignored, the variables Ω_j^* will behave as *reaction/interface transfer invariants*. The same behaviour will hold in the present system, but under the additional assumption that the fluxes of the transferred species on the walls can still be correctly evaluated. For an explicit visualisation of such a behaviour of the Ω_j^*, their governing equations obtained from Eq. (3.44a–d) by making similar combinations as explained in Sect. 3.4.2.4 are written below as Eqs. (3.46a–c).

For steady state, common dispersion coefficient D and $j > K + E$:

$$(1 - \Omega_E)[\underline{G} \cdot \nabla\Omega_j^* - \nabla \cdot (\rho D \nabla\Omega_j^*)] = -2\rho D \nabla\Omega_E \cdot \nabla\Omega_j^* \tag{3.46a}$$

Boundary conditions:

$$\begin{aligned} &(1 - \Omega_{E0})G_0[\Omega_j^*(\underline{z}_0) - \Omega_{j0}^*] = \left[(1 - \Omega_E)\rho D \nabla_n \Omega_j^*\right]_{\underline{z}_0}; \\ &(\nabla_n \Omega_j^*)_{\underline{z}_s} = 0 \end{aligned} \tag{3.46b}$$

$$-[(1 - \Omega_E)\rho D \nabla_n \Omega_j^*]_w = \mathbb{G}_{jw} - \Omega_j^*(\underline{z}_w)(G_w - \mathbb{G}_{Ew}) \tag{3.46c}$$

When only species A_e are transferred, $\mathbb{G}_{jw} = 0$ and $G_w = \mathbb{G}_{Ew}$. The right-hand side of Eq. (3.46c) becomes zero and the system (3.46a–c) is satisfied with $\Omega_j^* = \Omega_{j0}^*$.

If the use of a common value D and/or the assumption that strictly $\mathbb{G}_{jw} = 0$ ($j > K + E$) are not considered as being suitable, the resolution procedure of Sect. 3.6.2.2 can be applied, using Eqs. (3.44a–d) for variables Ω_j in the turbulent region and following the application guidelines discussed at the end of Sect. 3.4.2.4.

3.7 Catalytic Packed-Bed Reactors in More Than One Spatial Coordinate

It is considered in this section fixed-bed reactors packed with catalytic porous pellets and single-phase flow. A model accounting for variable changes in two or three spatial coordinates will be employed. Such a model conceives variables averaged on a volume control, whose dimensions are in the order of particle size. This kind of models are sometimes termed "*effective medium model*". The conservation equations of such models should rely upon effective parameters for quantifying the dispersive transport of mass and heat (*effective diffusivities and conductivities*) and the mass and energy interfacial exchange (*mass and heat transfer coefficients*).

The model presented here includes the distribution of volume averaged velocities in the domain, which depends on the bed porosity ε and its variations in regions close to the walls. The velocity distribution can be evaluated by the use of some version of the Brinkman's equation that accounts for momentum transfer from the fluid to the packing (see e.g., Luzi et al. [16]).

The evaluation of the catalytic reaction rates strictly needs for the solution of the species and energy conservation equations inside the pellets. For the purpose in this section, it will suffice to formulate global mass balances and to retain the conclusions in Sect. 3.5 about the role of the component variables Ω_j in a catalytic slab.

Let $\underline{u}$ and $\underline{G} = \rho\underline{u}$ be the *superficial* velocities and mass-velocities vectors. Then, the mass-flux vector of species A_j is expressed as:

$$\underline{G}_j = \omega_j \underline{G} + \underline{j}_j \tag{3.47}$$

where the dispersive mass-flux vector is $\underline{j}_j = -\rho D_j \nabla \omega_j$, and D_j is a spatial tensor of effective diffusivities, which can be expressed by adding molecular and dispersive contributions:

$$D_j = (\varepsilon/\tau) I \mathfrak{D}_j + D_{j,conv} \tag{3.48}$$

where I is the unit tensor, $\mathfrak{D}_j$ is the molecular diffusivity of A_j in the mixture (e.g., evaluated as in Eq. 3.16, τ is the tortuosity factor, and $D_{j,conv}$ is the tensor of *dispersion coefficients* originated by the irregular nature of the flow in packed beds at the particle-size scale: the fluid velocity shows sharp profiles between the surfaces of the pellets and frequent deflections because of the tortuous interstitial paths.

$D_{j,conv}$ is expressed as a tensor because the dispersivity in the net direction of fluid flow is several times larger than in the transversal directions. The components of $D_{j,conv}$

will usually be significantly higher than the molecular diffusivities $\mathfrak{D}_j$ at molecular Peclet numbers $Pe_{m,j} = Gd_p/(\rho\mathfrak{D}_j)$ larger than about 10, frequently found in practical applications.

The principal values of $D_{j,conv}$ can be evaluated from correlations depending on the particle Reynolds number $Re_p = Gd_p/\mu$ and Schmidt number $Sc_j = \mu/(\rho\mathfrak{D}_j)$ (note that $Pe_{m,j} = Re_p Sc_j$). The effect of Sc_j is moderate for liquids, low for gases and negligible at high values of Re_p (see e.g., [17]). It is noted that correlations for $D_{j,conv}$ do not customarily discriminate the flow regime in packed beds (Darcy, Forchheimer, transitional or turbulent).

Due to the usual dominance of $D_{j,conv}$ in Eq. (3.48), D_j will only depend slightly on the species identity, and the assumption of a common tensor $D_j = D$ will be satisfactory for moderately high values of Re_p.

The total mass conservation equation in the fluid phase is expressed as:

$$\varepsilon\frac{\partial\rho}{\partial t} + \nabla\cdot\underline{G} = -(1-\varepsilon)a_pG_p, \tag{3.49a}$$

where a_p is the surface-area to volume ratio of the pellet and G_p is the total mass flux towards the pellets.

For simplicity, it will be assumed here that the density inside the pellet pores and in the surrounding fluid phase is the same. This assumption will not alter the subsequent analysis. Then, a total mass balance in the fluid inside the particles can be written as:

$$(1-\varepsilon)\varepsilon_p\frac{\partial\rho}{\partial t} = (1-\varepsilon)a_pG_p, \tag{3.49b}$$

where it is clear that G_p will only be non-zero in a transient state when temporal density changes should be considered.

By adding Eqs. (3.49a, b), the total mass balance in the composite medium arises:

$$\left[\varepsilon + (1-\varepsilon)\varepsilon_p\right]\frac{\partial\rho}{\partial t} + \nabla\cdot\underline{G} = 0 \tag{3.49c}$$

The species conservation equations in the fluid phase are:

$$\varepsilon\frac{\partial(\rho\omega_j)}{\partial t} + \nabla\cdot\underline{G}_j = -(1-\varepsilon)a_pG_{jp}, \tag{3.50a}$$

where G_{jp} is the mass flux of species A_j towards the pellets.

A global balance for the species A_j inside a pellet is expressed as:

$$(1-\varepsilon)\varepsilon_p\frac{\partial(\rho\overline{\omega}_{jp})}{\partial t} - (1-\varepsilon)\overline{\hat{r}}_j = (1-\varepsilon)a_pG_{jp}, \tag{3.50b}$$

where $\overline{\omega}_{jp}$ is the average mass fraction of A_j inside the pellet and $\overline{\hat{r}}_j$ is the average net generation rate of species A_j per unit pellet volume. Equation (3.50b) neglects the mass exchange rate between neighbouring pellets, as is much lower than the pellet–fluid exchange rate G_{jp}.

Making the operation between equations [(3.50*a*) + (3.50*b*) − $\omega_j \times$ (3.49*a*) − $\overline{\omega}_{jp} \times$ (3.49*b*)] and using Eq. (3.47) for $\underline{G}_j$, the species balances in the composite medium are obtained:

$$\varepsilon\rho\frac{\partial\omega_j}{\partial t} + (1-\varepsilon)\varepsilon_p\rho\frac{\partial\overline{\omega}_{jp}}{\partial t} + \underline{G}\cdot\nabla\omega_j - \nabla\cdot\left(\rho D_j\nabla\omega_j\right) = (1-\varepsilon)\left[\overline{\hat{r}}_j + a_pG_p(\omega_j - \overline{\omega}_{jp})\right] \tag{3.50c}$$

To formulate the governing equations for the component variables $\Omega_j = \omega_j - \sum_k \hat{\sigma}_{kj}\omega_k$ in the fluid, it should be considered that $\overline{\hat{r}}_j = \sum_k \hat{\sigma}_{kj}\overline{\hat{r}}_k$ is satisfied for the average catalytic reactions and the average component variables inside the pellets are defined as $\overline{\Omega}_{jp} = \overline{\omega}_{jp} - \sum_k \hat{\sigma}_{kj}\overline{\omega}_{kp}$. Then, the equation set (3.50c) can be split in a subset for the key species mass fractions ω_k and another for the component variables Ω_j, by making the equivalent algebraic combinations as made in Sect. 3.6:

Key Species, $k = 1, ..., K$:

$$\varepsilon\rho\frac{\partial\omega_k}{\partial t} + (1-\varepsilon)\varepsilon_p\rho\frac{\partial\overline{\omega}_{kp}}{\partial t} + \underline{G}\cdot\nabla\omega_k - \nabla\cdot(\rho D_k\nabla\omega_k) = (1-\varepsilon)\left[\overline{\hat{r}}_k + a_pG_p(\omega_k - \overline{\omega}_{kp})\right] \tag{3.51}$$

Component Species, $j > K$:

$$\varepsilon\rho\frac{\partial\Omega_j}{\partial t} + (1-\varepsilon)\varepsilon_p\rho\frac{\partial\overline{\Omega}_{jp}}{\partial t} + \underline{G}\cdot\nabla\Omega_j - \nabla\cdot\left(\rho D_j\nabla\Omega_j\right) - \nabla\cdot[\rho\sum_k\hat{\sigma}_{kj}\left(D_j - D_k\right)\nabla\omega_k] = (1-\varepsilon)a_pG_p(\Omega_j - \overline{\Omega}_{jp}) \tag{3.52}$$

Boundary and initial conditions for the model in this section are assumed the same as those considered in Sect. 3.5 and are identically formulated. In general, permeable walls are also considered. For the vector $\underline{G}$, the entry condition $G_n(\underline{z}_0) = G_0$ (Eq. 3.39b) and Eq. (3.40) on the wall apply. For the mass fractions ω_k and component variables Ω_j, Eqs. (3.43b–d) and (3.44b–d) hold, respectively.

In the usual case of diffusion limitations inside the catalytic pellets, the reaction rates $\overline{\hat{r}}_k$ in Eq. (3.51) should be evaluated by solving the governing equations inside the pellets in a similar way as discussed in the example of Case I in Sect. 3.5 and in Appendix 3, where external mass transfer limitations and transient behaviour are considered.

It is also important to recall that homogeneous reactions can occur simultaneously with catalytic reactions, usually as side reactions. The formers will take place in the fluid stream and inside the pellet pores. A single set of both, homogeneous and catalytic reactions, should be considered and key/component species chosen accordingly. In Eq. (3.51), the net production rates of the homogeneous reactions in the fluid stream and in the pellet pores should be added, while Eq. (3.52) for the Ω_j remain the same. For the discussion in the next section, no specific feature is introduced by the appearance of homogeneous reactions.

3.7.1 Significance of Component Variables Ω_j in Catalytic Packed-Bed Reactors

The conservation equations and transport properties for the present problem can be compared with those in Sect. 3.6 for the homogeneous reactor with turbulent flow. Concerning the effective diffusivities in Eq. (3.48), the impact of the molecular contributions $\mathfrak{D}_j$ is only significant up to very short distances from the walls, less than a particle radius. At longer distances, the fluid velocities reach values comparable to that in the bulk of the bed and $D_{j,conv}$ becomes dominant. Hence, it will be assumed in this section that only small differences in the D_j may arise. In addition, it will be normally precise enough to extend the behaviour of the bulk of the bed to the whole cross-section. Regarding the use of variables Ω_j, this simplification removes the need of dividing the domain for the application of the resolution procedure introduced in Sect. 3.4.2.2, as discussed in Sect. 3.6.2.2 for the turbulent flow systems.

In steady state, Eqs. (3.51) and (3.52) are formally the same as Eqs. (3.43a) and (3.44a), respectively, and the role of the variables Ω_j will be similar in both systems, except for the wall region in the homogeneous reactor with turbulent flow.

In a transient state, the accumulation terms taking place inside the pellets, $(\partial\overline{\omega}_{jp}/\partial t)$ and $G_p(\omega_j - \overline{\omega}_{jp})$ for the key species, and $(\partial\overline{\Omega}_{jp}/\partial t)$ and $G_p(\Omega_j - \overline{\Omega}_{jp})$ for the component variables, have no counterpart in Eqs. (3.43a) and (3.44a). The term $G_p(\Omega_j - \overline{\Omega}_{jp})$ will not be particularly significant, except for large changes in density. Instead, the term $(\partial\overline{\Omega}_{jp}/\partial t)$ will be as significant as the variation $(\partial\Omega_j/\partial t)$, but the average value $\overline{\Omega}_{jp}$ will in general depend on the composition inside the particle (and hence on reaction effects), as follows from the conclusions reached in Sect. 3.5. This dependence will be inherited by variables Ω_j in the fluid stream. Irrespective of the role of variables Ω_j, it should be mentioned that in a transient state the major calculation burden corresponds to the solution of the microscopic conservation balances inside the particles, which should be made in each node of a spatial grid spanning the reactor domain.

To alleviate such a computational burden, the use of the *pseudo-steady state hypothesis* (PSSH) is frequently suggested, meaning the evaluation of the rates $\overline{\hat{r}}_j$ by ignoring any effect of mass accumulation inside the pellets. In this way, the equivalence between Eqs. (3.43a, 3.44a) and (3.51, 3.52) in a transient state is restored. The PSSH can be reasonably tried when, in the time scale desirable to track the evolution of the system, the accumulation rates of the species inside the catalyst happen to be considerably less than their net consumption rates. In cases when the catalytic reactions lead to multiplicity of steady states inside the particles, the use of the PSSH will not allow predicting the actual evolution of the system.

The particular cases for the convenient use of variables Ω_j in catalytic packed-bed reactors can be summarised with reference to Sect. 3.6.2 and taking into account the previous discussion about similarities and differences with the homogeneous turbulent-flow system. When the resolution procedure introduced in Sect. 3.4.2.2 can be used, it is recalled that it will be applied for the whole domain. It follows that:

- If the PSSH is employed, all the particular cases described in Sects. 3.6.2.1, 3.6.2.3 and 3.6.2.4 will also apply for catalytic packed-bed reactors.
- If the accumulation terms in the pellets are retained in a transient state, the particular cases described in Sects. 3.6.2.1 and 3.6.2.3 no longer apply.

3.8 Adiabatic Operations in Steady State

It was briefly discussed in Sect. 2.5.5 of Chap. 2 that for an adiabatic plug-flow reactor in steady state the temperature changes ($T - T_0$) in any position could be related with the progress of the reaction rates, as expressed by the differences ($y_k - y_{k0}$) of the key species (Eq. 2.61 in Chap. 2). This relationship is actually a reduced form of the thermal energy balance, and it facilitates the solution of all governing equations. Although, it was not emphasised in Chap. 2, such a relationship requires that the specific enthalpy $\hat{h}$ of the mixture remains constant, a condition that is reasonably satisfied at low pressure drops and negligible viscous dissipation effects. In general, these assumptions can be safely made for flow reactors (see e.g., Froment et al. [18]).

Since adiabatic processes are frequent in practice, this section is dedicated to identifying conditions leading to maintaining $\hat{h}$ constant in the whole domain of the models in Sects. 3.6 and 3.7, to finally relate temperature changes ($T - T_0$) with the key species changes ($\omega_k - \omega_{k0}$). Possible situations to that end are restricted to the case of impermeable walls and steady-state operations. It is anticipated that restrictions will be imposed by mass and heat dispersion effects considered in the models of Sect. 3.6 and 3.7. The results will be also applicable to the ADM in Sect. 3.4, as it can be regarded as a 1D version of the models in Sects. 3.6 and 3.7.

The expressions for the specific enthalpy $\hat{h}$ and its variations with temperature and composition are first recalled. In terms of the specific enthalpy of the species $\hat{h}_j$ (partial molar enthalpy divided by m_j):

$$\hat{h} = \sum_j \omega_j \hat{h}_j \tag{3.53}$$

Spatial variations of $\hat{h}$ can be written as:

$$\nabla\hat{h} = \sum_j \omega_j \nabla\hat{h}_j + \sum_j \hat{h}_j \nabla\omega_j \tag{3.54a}$$

Recalling that the effect of pressure is neglected, $\nabla \hat{h}_j$ is expressed as:

$$\nabla \hat{h}_j = \hat{c}_{pj} \nabla T + \nabla(\hat{h}_j)_{T,P} \tag{3.54b}$$

where $\hat{c}_{pj}$ is the specific heat capacity at constant pressure of species A_j and $\nabla(\hat{h}_j)_{T,P}$ accounts for the changes of $\hat{h}_j$ caused by composition variations. Replacing $\nabla \hat{h}_j$ in Eq. (3.54a),

$$\nabla \hat{h} = \hat{c}_p \nabla T + \sum_j \omega_j \nabla(\hat{h}_j)_{T,P} + \sum_j \hat{h}_j \nabla \omega_j \tag{3.54c}$$

where $\hat{c}_p$ is the specific heat capacity of the mixture at constant pressure:

$$\hat{c}_p = \sum_j \omega_j \hat{c}_{pj} \tag{3.55}$$

The Gibbs–Duhem expression establishes that $\sum_j \omega_j \nabla(\hat{h}_j)_{T,P} = 0$. Therefore, Eq. (3.54c) becomes:

$$\nabla \hat{h} = \hat{c}_p \nabla T + \sum_j \hat{h}_j \nabla \omega_j \tag{3.56}$$

3.8.1 Homogeneous Reactors with Turbulent Flow

For the model considered in Sect. 3.6, the dispersive heat-flux vector $\underline{\varphi}$ is defined as

$$\underline{\varphi} = -\Lambda \nabla T + \sum_j \hat{h}_j \underline{j}_j \tag{3.57}$$

where $\underline{j}_j = -\rho D_j \nabla \omega_j$ is the dispersion mass-flux vector of species A_j (Eqs. 3.37 and 3.38) and Λ is an effective conductivity adding the molecular thermal conductivity λ and the turbulent thermal conductivity Λ^t. In a similar way as D_j (Eq. 3.38), $\Lambda = \lambda + \Lambda^t$.

The local balance of thermal energy in steady state ($\nabla \cdot \underline{G} = 0$) is expressed as:

$$\underline{G} \cdot \nabla \hat{h} + \nabla \cdot \underline{\varphi} = 0 \tag{3.58}$$

where pressure changes and viscous dissipation effects have been neglected. Replacing $\underline{\varphi}$ from Eq. (3.57), it is obtained:

$$\underline{G} \cdot \nabla \hat{h} - \nabla \cdot [\Lambda \nabla T + \rho \sum_j (\hat{h}_j D_j \nabla \omega_j)] = 0 \tag{3.59}$$

Replacing ∇T from Eq. (3.56) and rearranging:

$$\underline{G} \cdot \nabla \hat{h} - \nabla \cdot [(\Lambda/\hat{c}_p)\nabla \hat{h}] - \nabla \cdot \left[\sum_j (\rho D_j - \Lambda/\hat{c}_p)\hat{h}_j \nabla \omega_j \right] = 0 \tag{3.60}$$

For $\hat{h}$ to remain constant it should be $\nabla \hat{h} = 0$ and this is feasible if the third term in Eq. (3.60) is zero. This requirement can be in general accomplished if each of the coefficient $(\rho D_j - \Lambda/\hat{c}_p)$ is zero. In turn, it should be verified that:

$$D_j = D \tag{3.61a}$$

$$\Lambda = \rho \hat{c}_p D \tag{3.61b}$$

It was discussed in Sect. 3.6.2.1 that the condition (3.61a) can be reasonably adopted in the whole domain for high enough Reynolds numbers, $D = D^t$. It is recalled that in the SS and when the walls are impermeable, the Ω_j will behave as reaction invariants, $\Omega_j = \Omega_{j0}$, if $D_j = D = D^t$ (a fact that will be used in Sect. 3.8.3 for establishing a temperature–composition relationship). Simultaneously, the turbulent component of the effective conductivity Λ will also be dominant, $\Lambda = \Lambda^t$, and the analogy between turbulent mass and heat transport can be applied $\Lambda^t/(\rho \hat{c}_p) = D^t$, where $\Lambda^t/(\rho \hat{c}_p)$ is the *turbulent thermal diffusivity*. In this way, the condition (3.61b) is also satisfied.

By replacing $\underline{j}_j = -\rho D \nabla \omega_j$ in Eq. (3.57) and combining the result with Eq. (3.56), it is obtained:

$$(\Lambda = \rho \hat{c}_p D): \quad \underline{\varphi} = -(\Lambda/\hat{c}_p)\nabla \hat{h} \tag{3.62}$$

The heat balance (3.60) becomes:

$$(\Lambda = \rho \hat{c}_p D): \quad \underline{G} \cdot \nabla \hat{h} - \nabla \cdot [(\Lambda/\hat{c}_p)\nabla \hat{h}] = 0 \tag{3.63a}$$

As required, Eq. (3.63a) is satisfied when $\hat{h}$ is constant, which also implies that the dispersive heat-flux vector is identically null, $\underline{\varphi} = 0$. The boundary conditions should also be verified. On the walls, recalling the adiabatic and impermeable walls conditions, $(\nabla_n T)_w = 0$ and $(\nabla_n \omega_j)_w = 0$. Then, from Eq. (3.56):

$$(\nabla_n \hat{h})_w = 0 \tag{3.63b}$$

At the exit section, the Danckwerts' conditions $(\nabla_n T)_{\underline{z}_s} = 0$ and $(\nabla_n \omega_j)_{\underline{z}_s} = 0$ lead to:

$$(\nabla_n \hat{h})_{\underline{z}_s} = 0 \tag{3.63c}$$

At the entry section, the energy balance written according to Danckwerts' conditions is:

$$(G_0\hat{h} + \varphi_n)_{\underline{z}_0} = G_0\hat{h}_0$$

From Eq. (3.62) for $\underline{\varphi}$:

$$(\Lambda = \rho\hat{c}_p D): \quad \left[G_0(\hat{h} - \hat{h}_0) - (\Lambda/\hat{c}_p)\nabla_n\hat{h}\right]_{\underline{z}_0} = 0 \tag{3.63d}$$

Equations (3.63a–d) are satisfied with the uniform value $\hat{h} = \hat{h}_0$.

3.8.2 Catalytic Packed-Bed Reactors

It was emphasised in Sect. 3.7 that the mass conservation equations for the model of catalytic packed-bed reactors in SS present the same structure as those of the homogeneous reactor with turbulent flow. If the corresponding energy balances also show the same similarity under the assumptions here considered, restriction as those in Eqs. (3.61a, b) will lead to maintaining $\hat{h}$ uniform.

In terms of variables averaged on a control volume, the steady-state energy balance in the composite medium of catalytic packed-bed reactors can be formally expressed (with a suitable definition of Λ) as in Eq. (3.59), provided that pellet-to-pellet heat and mass exchange can be disregarded. The pellet-to-pellet mass exchange can be neglected in the vast majority of practical situations, as mentioned in Sect. 3.7 (see Eq. 3.50b). Instead, the pellet-to-pellet heat exchange can be significant, due to the normally high solid thermal conductivities. Such a heat exchange takes place by a series conductive mechanism involving the thin fluid-fillets interposed between the pellets. The resulting heat flux, on a superficial basis, can be expressed as $\underline{\varphi}_p = -\Lambda_{p,ef}\nabla T_p$, where $\Lambda_{p,ef}$ is an effective pellet-to-pellet thermal conductivity and ∇T_p is the average temperature gradient through the pellets. This mechanism is important for transversal heat transport to the wall in non-adiabatic processes, when the transversal component of ∇T_p can be different to that in the fluid, particularly close to the heat transfer wall. Instead, in adiabatic operations, it can be safely assumed that $\nabla T_p = \nabla T$. In this way, by adding $\underline{\varphi}_p$ in the energy balance, Eq. (3.59) will be formally valid if Λ is defined as:

$$\Lambda = \Lambda_{conv} + I[(\varepsilon/\tau)\lambda] + I\Lambda_{p,ef} \tag{3.64}$$

where $\Lambda_{conv} + I[(\varepsilon/\tau)\lambda]$ corresponds to the effective conductivity tensor in the interstitial fluid, expressed similarly as the effective diffusivity tensor (Eq. 3.48), and Λ_{conv} is the tensor of *thermal dispersion coefficients*, arisen from similar mechanisms as $D_{j,conv}$ (Sect. 3.7).

Therefore, the analysis made in Sect. 3.8.1 to disclose sufficient conditions to maintain $\hat{h}$ uniform follows in the same way; hence, the equalities (3.61a, b) will be needed to hold. Equation (3.61a) will be satisfied at relatively high values of the particle Reynolds number Re_p, $D_j = D = D_{conv}$, as discussed in Sect. 3.7. Then, the Ω_j will be reaction invariants $\Omega_j = \Omega_{j0}$ (Sect. 3.8.1). As regards condition (3.61b), it is first noted that Λ_{conv} will be normally higher than the molecular contribution $I[(\varepsilon/\tau)\lambda]$ and that the analogy $\Lambda_{conv} = \rho\hat{c}_p D_{conv}$ will simultaneously apply when $D_j = D_{conv}$. Nonetheless, Λ_{conv} still has to be significantly larger than $I\Lambda_{p,ef}$ (Eq. 3.64) to fulfil Eq. (3.61b), a fact that eventually will need for higher values of Re_p.

In conclusion, it is appraised that Eqs. (3.61a, b) can be satisfied in many practical situations to maintain the specific enthalpy uniform in catalytic packed-bed reactors.

3.8.3 Relationship Between Temperature and Composition from $\hat{h} = \hat{h}_0$

Once the condition $\hat{h} = \hat{h}_0$ is satisfied, it can be written $\hat{h}(T, \omega_1, \ldots, \omega_S) = \hat{h}_0$. As simultaneously it was shown that $\Omega_j = \Omega_{j0}$, the ω_j for the component species can be written as $\omega_j = \omega_{j0} + \sum_k \hat{\sigma}_{kj}(\omega_k - \omega_{k0})$, and hence it follows that $\hat{h}(T, \omega_1, \ldots, \omega_K) = \hat{h}_0$. This expression allows relating at any position in the system the values of T and ω_k, for known input values T_0, ω_{j0}. However, an explicit relationship is desirable, although inevitably at the cost of some loss of accuracy. To this end, an equivalent differential form of Eq. (3.56) with $\hat{h} = \hat{h}_0$ can be written:

$$\sum_j \hat{h}_j d\omega_j + \hat{c}_p dT = 0 \tag{3.65}$$

Using $d\omega_j = \sum_k \hat{\sigma}_{kj} d\omega_k$ for the component species in Eq. (3.65):

$$\sum_k \Delta\hat{H}_k d\omega_k + \hat{c}_p dT = 0 \tag{3.66}$$

where the *specific reaction enthalpy* $\Delta\hat{H}_k$ of the canonical reaction $\mathcal{R}_k$ is defined as:

$$\Delta\hat{H}_k = \hat{h}_k + \sum_j \hat{\sigma}_{kj}\hat{h}_j \tag{3.67}$$

To proceed, the customary assumption is that the ratio $(\Delta\hat{H}_k/\hat{c}_p)$ in Eq. (3.66) remains constant. Then, for values $(\Delta\hat{H}_k/\hat{c}_p)_{ref}$ evaluated at certain reference "*ref*" conditions, the integration of (3.66) is trivial:

$$T - T_0 = \sum_k (-\Delta\hat{H}_k/\hat{c}_p)_{ref}(\omega_k - \omega_{k0}) \tag{3.68}$$

Equation (3.68) is the same kind of relationship employed for adiabatic reactors in introductory CRE courses, as shown in Chap. 2. In addition to the simplification gained in accounting for temperature changes, Eq. (3.68) provides an important conceptual plus. The simplest choice for the reference conditions is clearly the evaluation at input conditions, $(-\Delta\hat{H}_k/\hat{c}_p)_{ref} = (-\Delta\hat{H}_k/\hat{c}_p)_0$.

The evaluation of temperature from Eq. (3.68) may not be accurate enough, particularly when rather large temperature changes take place and cause significant variations of $(\Delta\hat{H}_k/\hat{c}_p)$. This feature will be inconsistent with the use of detailed models, as in Sects. 3.6 or 3.7. In such cases, Eq. (3.66) is suitable to be introduced -as the thermal energy balance- in the numerical procedure to resolve the remainder governing equations (as described in Sects. 3.6 and 3.7). Values of $(\Delta\hat{H}_k/\hat{c}_p)$ can thus be evaluated at local values of T and ω_k.

As a final remark, it is recalled that in Sects. 3.8.1 and 3.8.2 the requirement of high enough Reynolds numbers (Re or Re_p) is needed to fulfil the restrictions expressed by Eqs. (3.61a, b). In practice, as the effects of mass and heat dispersion in adiabatic flow reactors are usually of secondary importance, the assumption of uniform $\hat{h}$, leading to the relevant Eq. (3.66) or (3.68), can be satisfactory even under conditions when Eq. (3.61a, b) do not strictly hold.

3.9 Significance of Variables Ω_j in Other Reacting Flow Systems or Models

Catalysed-wall reactors, especially monolithic reactors [19], operate normally in laminar regime. If a 1D model is employed to evaluate the axial changes of the state variables, the exchange rate of the species with the catalytic wall is evaluated by employing mass transfer coefficients. The structure of the governing equations of this model will resemble a catalytic packed-bed reactor with plug flow, i.e., a simplified version of the model in Sect. 3.7. As the catalytic layer on the wall (*washcoat*) is very thin, the species accumulation therein can be neglected, and the pseudo-steady state hypothesis (PSSH) (Sect. 3.7.1) can be applied. Therefore, the use of the variables Ω_j will bring about similar benefits as those discussed in Sects. 3.4.2 and 3.7.1.

Instead, a 3D model (typical cross-sections of monolithic channels are not circular) can be employed to avoid the use of mass transfer coefficients. The transversal diffusive fluxes of the species will be relevant over the whole cross-section. As a consequence, significant differences between the molecular diffusivities $\mathfrak{D}_j$ will prevent variables Ω_j from offering a computational advantage, as discussed conceptually in *Case II* of Sect. 3.5.

Given the increasing application of CFD techniques in chemical reaction systems during the last two decades, it is important to reveal the significance of variables Ω_j in such a context. The CFD approach is intended to resolve the governing equations of the state variables in a small enough scale to capture all their relevant changes. Of particular importance here is that interfacial mass transfer can be predicted without resorting to empirical mass transfer coefficients.

The 3D model of monolithic reactors may be regarded as a CFD application for a laminar flow system. Also, the governing equations of the turbulent flow systems in Sect. 3.5 are similar to those from a CFD application. In these cases, the limitations imposed by molecular diffusion on the use of variables Ω_j have been already emphasised.

The use of CFD techniques for catalytic packed-bed reactors should be finally commented on. It is first noted that the term "CFD approach" is employed in some studies for the resolution of *effective medium models* based on volume averaged variables and effective transport properties, as mentioned in Sect. 3.7. In these cases, the discussion in Sect. 3.7.1 to assess the relevance of variables Ω_j applies. Instead, the term *particle resolved CFD simulations* (PRCFDS) is frequently employed when the field of state variables in the whole volume of the interstitial fluid is evaluated with due account of the solid-surface boundaries [20]. Then, this approach allows evaluating local rates of mass exchange with the catalytic pellets through the boundary layer on their surfaces, where again molecular diffusion in the directions normal to the pellet surface is most relevant and, therefore, variables Ω_j cannot be conveniently used (*Case II* of Sect. 3.5). Consequently, only in the interstitial regions sufficiently apart from the pellet surfaces, the variables Ω_j may behave nearly independent of the state variables. Although a resolution strategy similar to that explained in Sect. 3.6.2.2 can be envisaged, discriminating regions "*close to the pellet surfaces*" and "*far from the pellet surfaces*", the definition of the boundaries between them would be a very difficult task, because of the complex and irregular distribution of the pellets in the bed. A dynamical definition of the boundaries during the resolution process could be a feasible alternative, but clearly at the expense of developing a more sophisticated resolution strategy. Summing up, the use of variables Ω_j does not seem to be particularly attractive for PRCFDS.

3.10 Final Remarks

For the presentation of this chapter, it was first recalled that mass fractions have been found suitable to define the composition of a flow reaction system. Accordingly, the reaction stoichiometry is expressed in mass, rather than molar, terms. In particular for the canonical reactions, the stoichiometric relationships between the net production rates of component and key species are written as $\hat{r}_j = \sum_k \hat{\sigma}_{kj} \hat{r}_k$ (Sect. 3.2).

The set of balance equations of the species in a single-phase reaction system can be always split in a subset for the key species and another subset for the so-called *component variables* (Sect. 3.3) that excludes the appearance of reaction rates. Each component variable combines a variable accounting for the mass present of a given component species

A_j with the equivalent variables of all key species weighted by the stoichiometric coefficients $\hat{\sigma}_{kj}$. For flow reaction systems undertaken in this chapter, the component variables $\Omega_j = \omega_j - \sum_k \hat{\sigma}_{kj} \omega_k$ $(j > K)$ are the best suited for that purpose.

It is expected that such decomposition will allow carrying out an independent solution of the subset for the component variables and afterwards proceeding with the solution of the key-species subset. If proved to be valid, the component variables are said to be *reaction invariants*. The procedure will avoid the simultaneous solution of the whole set and, therefore, reduce the computational burden. Simultaneously, it can be concluded that the mass fractions of the key species will be the only variables needed to establish the composition of the mixture at each position in the system.

A sufficient condition for variables Ω_j to behave as reaction invariants is that the resolution of their governing equations does not involve any property that depends on the state-variables (i.e., composition, temperature, pressure), since these will in turn depend on the reaction rates $\hat{r}_k$ (Sect. 3.4.1).

This condition is satisfied for the plug-flow model of a tubular reactor in steady state (Sect. 3.3). However, when the effect of the species dispersion is considered, as in the Axial Dispersion model (ADM), dispersion coefficients D_j depending on the species identity will prevent variables Ω_j to behave as reaction invariants. Moreover, transient operations and the possibility for mass exchange through the vessel wall were considered for the ADM analysed in Sect. 3.4. In general, any of these added features will also prevent the variables Ω_j from being reaction invariant. The models in more than one spatial coordinate treated in Sect. 3.6 for homogeneous systems under turbulent regime and in Sect. 3.7 for catalytic packed-bed reactors present analogous limitations.

In view of the above general restrictions, the primary objective in this chapter was to identify assumptions of practical significance that allow relaxing such restrictions.

Considering specifically the ADM, in the frequent case of impermeable walls and steady state, a high enough Reynolds number Re allows assuming a single dispersion coefficient D for all the species and consequently the behaviour of variables Ω_j as reaction invariants (Sect. 3.4.2.1). The same conclusion holds under transient condition if constant values of fluid density and coefficient D can be assumed. For operations with mass exchange through the walls, if only some of the component species are involved, for the remainder component species the modified component variables $\Omega_j^* = \Omega_j/(1 - \Omega_E)$ (Ω_E is the sum of Ω_j of the selectively transferred species) behave as reaction invariants in steady state (Sect. 3.4.2.4).

In the case that the above-explained assumptions cannot be strictly sustained, variables Ω_j may still be used to reduce the computational demand in the process of solving the governing equations, provided that the Ω_j will depart moderately from the behaviour as reaction invariants. Considering the normally necessary application of an iterative procedure for solving simultaneously the whole set of non-linear equations, it was proposed to iterate alternatively on the subset of Ω_j-equations and on the subset for the remainder equations (which include the key-species equations). Thus, the subset of Ω_j-equations becomes linear and uncoupled and therefore updated values of Ω_j can be readily evaluated, while the dimensionality of the remainder set can be noticeably reduced (Sects. 3.4.2.2, 3.4.2.3 and 3.4.2.4).

The model in more than one spatial coordinate analysed in Sect. 3.6 aggravates the effect of different dispersion coefficients D_j on the behaviour of the variables Ω_j. This is due to the fact that in the transversal directions to the global flow, the total flux of the species is governed by dispersion. Even under the assumed turbulent regime, molecular diffusion is the main mechanism for the species transport at positions close enough to the vessel walls. As the molecular diffusivities $\mathfrak{D}_j$ are likely to depend significantly on the species identity, the variables Ω_j can also lose practical significance, as discussed specifically in Sect. 3.5. However, the size of the "wall region" under the strong influence of the $\mathfrak{D}_j$ and its effects can be neglected at sufficiently high Reynolds numbers and a common dispersion coefficient D can be assumed for the whole domain. Under these conditions, the Ω_j or the Ω_j^* can behave as reactions invariants (Sects. 3.6.2.1 and 3.6.2.4), similarly as for the ADM. Otherwise, if the size of the wall region is significant, the proposed alternate-iteration procedure mentioned above can be used in the "turbulent region" and the simultaneous iteration of the whole equation set carried out in the wall region (Sects. 3.6.2.2–3.5).

An *effective medium model* was used in Sect. 3.7 to analyse catalytic packed-bed reactors. The mass conservation equations and their behaviour with respect to the variables Ω_j are similar to those in the case of the homogeneous system in Sect. 3.6, except for two different features. The size of the "wall region", where molecular diffusion is relevant, is very thin and its effect can be neglected. Then, dispersion is dominated by convection effects and the alternate-iteration procedure can be employed in the whole domain. The second different feature is found in a transient operation. The effective diffusivities inside the pellets behave similarly as molecular diffusivities with respect to the dependency on the species identity. Hence, the variables Ω_j are not suitable to evaluate the accumulation of the species inside the catalytic pellets (Sect. 3.5). In general, this effect prevents the Ω_j from a beneficial usage in transient conditions. Instead, if the accumulation inside the pellets can be neglected (pseudo-steady state hypothesis), the significance of the variables Ω_j becomes equivalent as for the homogeneous system in Sect. 3.6.

Adiabatic processes in steady state are commonly found in practice. For the simple plug-flow model the specific enthalpy of the mixture $\hat{h}$ remains constant all along the vessel. This condition allows reducing the thermal energy balance to a relationship between temperature $(T - T_0)$ and key-species $(\omega_k - \omega_{k0})$ changes, which greatly facilitates the resolution of governing equations. For the systems in this chapter with impermeable walls (Sects. 3.4, 3.6 and 3.7), it was shown in Sect. 3.8 that $\hat{h}$ will remain strictly constant when the same value for the mass dispersion coefficients and for the effective thermal diffusivity can be assumed. This requirement will be practically satisfied in operations with high enough Reynolds numbers.

CFD modelling allows evaluating mass interfacial exchange, avoiding the use of mass transfer coefficients (Sect. 3.9). The role of molecular diffusion is highly relevant in this process and in certain systems an important volume of the fluid stream can be involved. Consequently, the beneficial use of variables Ω_j can be significantly impaired, following the conclusions in Sect. 3.5. A clear example in this sense is given in the modelling of monolithic reactors with (normally) laminar flow, where mass transfer to the catalytic walls of the channel involves practically the whole cross-section. In the CFD applications for catalytic packed-bed reactors, the volume of the interstitial regions effectively participating in the mass transfer process to the pellet surfaces can also be significant and of difficult identification.

In summary, the stoichiometry-based component variables, specifically variables Ω_j, can be employed to improve the resolution of the governing-equation set of reaction-system models, either by independent resolution of two split subsets (reaction-invariance behaviour of the component variables) or by alternate iterations of such subsets. It was shown that such approach can be used in a variety of flow reaction systems of practical significance modelled with different levels of detail. Moreover, the sources of effects impeding the practical use of the component variables have been clearly identified and can be useful to analyse the significance of component variables in reacting systems or models not explicitly discussed in this chapter.

Appendix 1: Use of Consistent Multi-component Dispersion or Diffusion Fluxes

It was discussed in Sect. 3.4 that when Fickian expressions are employed for multi-component dispersion or diffusion fluxes, the consistency condition $\sum_j j_j = 0$ is not satisfy for the one-dimensional fluxes in the ADM. More generally, for 3D vector fluxes $\underline{j}_j = -\rho D_j \nabla\omega_j$, the condition $\sum_j \underline{j}_j = \underline{0}$ is not verified either. Kee et al. [8] suggested a simple correction to the Fickian fluxes to satisfy the consistency condition. For general 3D fluxes and in terms of, e.g., D_j:

$$\underline{j}_j = -\rho D_j \nabla\omega_j + \omega_j \underline{J}_c; \qquad \underline{J}_c = \rho \sum_{j'=1}^{S} D_{j'} \nabla\omega_{j'}, \tag{3.69}$$

where $\underline{J}_c$ is the correction term. Eq. (3.69) clearly satisfies $\sum_j \underline{j}_j = \underline{0}$, and it also applies for diffusion fluxes by replacing D_j with $\mathfrak{D}_j$. It is further noted that $\underline{J}_c = \underline{0}$ for a common value $D_j = D$. This fact warranties that the correction term will not modify any of the situation at which the variables Ω_j are found to behave as reaction invariants.

As an example of how the governing equations of the species balances are modified when the corrected fluxes are employed, Eqs. (3.20a, b) for the key species and Eqs. (3.22a, b) for the component variables Ω_j are rewritten here, considering the unidirectional fluxes in the ADM model, i.e., $j_j = -\rho D_j \partial\omega_j/\partial z + \omega_j J_c$ and $J_c = \rho\sum_{j'=1}^{S} D_{j'}\partial\omega_{j'}/\partial z$:

For Eqs. (3.20a, b), $k = 1, \ldots, K$:

$$\rho\frac{\partial\omega_k}{\partial t} + G\frac{\partial\omega_k}{\partial z} - \frac{\partial}{\partial z}\left(\rho D_k\frac{\partial\omega_k}{\partial z}\right) + \frac{\partial(\omega_k J_c)}{\partial z} + a_w(G_{kw} - \omega_k G_w) = \hat{r}_k; \tag{3.70}$$

$$\begin{aligned} &\omega_k(0,z) = \omega_{kI}(z); \\ &G_0(\omega_k - \omega_{k0})_{z=0} + \left[-\rho D_k\tfrac{\partial\omega_k}{\partial z} + \omega_k J_c\right]_{z=0} = 0; \\ &\left(\tfrac{\partial\omega_k}{\partial z}\right)_{z=L} = 0 \end{aligned} \tag{3.71}$$

For Eqs. (3.22a, b), $j > K$:

$$\begin{aligned} &\rho\tfrac{\partial\Omega_j}{\partial t} + G\tfrac{\partial\Omega_j}{\partial z} - \tfrac{\partial}{\partial z}\left(\rho D_j\tfrac{\partial\Omega_j}{\partial z}\right) = \\ &\quad -\tfrac{\partial(\Omega_j J_c)}{\partial z} + \tfrac{\partial}{\partial z}\left[\rho\sum_k \hat{\sigma}_{kj}(D_j - D_k)\tfrac{\partial\omega_k}{\partial z}\right] - a_w(\mathbb{G}_{jw} - \Omega_j G_w); \end{aligned} \tag{3.72}$$

$$\begin{aligned} &\Omega_j(0,z) = \Omega_{jI}(z); \\ &G_0(\Omega_j - \Omega_{j0})_{z=0} + \left[-\rho D_j\tfrac{\partial\Omega_j}{\partial z} + \Omega_j J_c\right]_{z=0} = \left[\rho\sum_k \hat{\sigma}_{kj}(D_j - D_k)\tfrac{\partial\omega_k}{\partial z}\right]_{z=0}; \\ &\left(\tfrac{\partial\Omega_j}{\partial z}\right)_{z=L} = 0 \end{aligned} \tag{3.73}$$

If the corrected flux vectors $\underline{j}_j$, Eq. (3.69), is used for the 3D models considered in Sects. 3.6 and 3.7, similar additional terms as in Eqs. (3.70)–(3.73) will arise, but with $\underline{J}_c$ instead of J_c.

When Eqs. (3.70)–(3.73) are solved according to the resolution procedure discussed in Sect. 3.4.2.2, in an updating step on Eqs. (3.72) and (3.73) for the variables Ω_j, it will be convenient to evaluate the discretised correction flux $J_c = \rho\sum_{j'=1}^{S} D_{j'}\partial\omega_{j'}/\partial z$ with the set of values $\{\omega_{j'}\} \equiv \{\omega_j\}$ (for all A_j), as updated from the previous iteration step of the key species A_k. In this way, Eqs. (3.72) and (3.73) will remain linear and uncoupled for updating each Ω_j.

Appendix 2: Normalisation of Ω_j Values

It is presented here a way to normalise the updated values of Ω_j in the resolution procedure introduced in Sect. 3.4.2.2, according to the consistency conditions $\sum_{j>K}\Omega_j = 1$ and values $\omega_j \geq 0$ of the component species ($j > K$) for current values ω_k of the key species. The scheme can be adapted for the situation with interfacial mass transfer (Sect. 3.4.2.4), as shown at the end of this appendix.

The ADM (Sect. 3.4) is referenced, but the present scheme is also applicable in the use of the resolution procedure for the systems treated in Sects. 3.6 and 3.7.

The solution of Eqs. (3.22a, b) will satisfy $\sum_{j>K}\Omega_j = 1$ when the input and initial values, Ω_{j0} and Ω_{jl}, meet that condition and the dispersion fluxes verify $\sum_j j_j = 0$. However, this last condition is not fulfilled when Fickian expressions are used for the fluxes. Instead, if the correction presented in Appendix 1 is employed for the fluxes, such a condition will be satisfied. Moreover, inaccuracies of numerical methods of solution can also contribute to introduce deviations. Therefore, the numerical values Ω_j obtained from Eqs. (3.22a, b) can be conveniently normalised by application of the following steps.

First, for each Ω_j $(j > K)$ arisen from Eqs. (3.22a, b) in a certain position, the mass fraction ω_j in Eq. (3.74) is evaluated with the current values of ω_k.

$$\omega_j = \Omega_j + \sum_k \hat{\sigma}_{kj}\omega_k; \qquad j > K \tag{3.74}$$

Then, assuming that there are M component species for which $\omega_j < 0$, indexed by $j = m = K+1, \ldots, K+M$, the Ω_m are replaced by corrected values Ω_m^{corr}:

$$\Omega_m^{corr} = -\sum_k \hat{\sigma}_{kj}\omega_k; \qquad K+1 \le m \le K+M \tag{3.75}$$

It is noted that with Ω_m^{corr} from Eq. (3.75), the corresponding mass fractions ω_m take the lower limit $\omega_m^{corr} = 0$.

For the remainder component species, $j > K+M$, the corrected mass fractions are expressed as:

$$\omega_j^{corr} = \alpha(\Omega_j + \sum_k \hat{\sigma}_{kj}\omega_k); \qquad j > K+M \tag{3.76}$$

where α is a common coefficient evaluated from adding all equations (3.76) and the result made equal to $(1 - \sum_k \omega_k)$:

$$\alpha = \frac{1 - \sum_k \omega_k}{\sum_{j>K+M} (\Omega_j + \sum_k \hat{\sigma}_{kj}\omega_k)} \tag{3.77}$$

Finally, the values Ω_j^{corr} for $j > K+M$ are set as $\Omega_j^{corr} = \omega_j^{corr} - \sum_k \hat{\sigma}_{kj}\omega_k$, or by replacing ω_j^{corr} from Eq. (3.76):

$$\Omega_j^{corr} = \alpha\Omega_j - (1-\alpha)\sum_k \hat{\sigma}_{kj}\omega_k; \qquad j > K+M \tag{3.78}$$

By summing Eqs. (3.75) and (3.78), it is obtained $\sum_{j>K}\Omega_j^{corr} = 1$, as desired. The normalised values Ω_j^{corr} $(j > K)$ are then used for a new iteration step of the key-species mass fractions ω_k. Since in this step the Ω_j^{corr} will be kept invariable and $\sum_{j>K}\Omega_j^{corr} = 1$ implies that $\sum_j \omega_j = 1$, the updated values ω_k will satisfy this condition. Therefore, in the execution of this step, it should be only controlled that the condition $0 \leq \omega_j \leq 1$ is accomplished for all the species.

The use of the resolution procedure of Sect. 3.4.2.2 is suggested in Sect. 3.4.2.4 when only E component species A_e $(K + 1 \leq j = e \leq K + E)$ are transferred significantly through permeable walls. In such a situation, only the values Ω_j associated with the remainder component species, $j > K + E$, are updated in each iteration step, with specified values of ω_k and ω_e. These values can be normalised in a similar way as described above, i.e., imposing the non-negativity condition to ω_j $(j > K + E)$ and, in this case, $\sum_{j>K+E} \Omega_j^{corr} = 1 - \Omega_E$. The expressions (3.74)–(3.78) hold with K substituted by $(K + E)$ and (3.77) modified in the following way:

$$\alpha = \frac{1 - \sum_e \omega_e - \sum_k \omega_k}{\sum_{j>K+E+M} (\Omega_j + \sum_k \hat{\sigma}_{kj}\omega_k)} \tag{3.79}$$

Then, the modified component variables are calculated, $\Omega_j^* = \Omega_j^{corr}/(1 - \Omega_E)$, which satisfy $\sum_{j>K+E} \Omega_j^* = 1$ (Eq. 3.25d). A new updating step for ω_k and ω_e is executed (Sect. 3.4.2.4) and, as in the case of impermeable walls, it should only be controlled that the condition $0 \leq \omega_j \leq 1$ is satisfied for all the species.

Appendix 3: Catalytic Porous Slab. External Mass Transfer Resistances and Transient Behaviour

The problem of diffusion/reaction in a catalytic porous slab in Sect. 3.5 is extended here by considering the effects of external mass transfer resistances and transient behaviour. The assumption of invariable ρ and $\mathfrak{D}_{j,ef}$ will be maintained.

In a general transient state, the species governing equations inside the slab are:

$$\rho \mathfrak{D}_{j,ef} \frac{\partial^2 \omega_j}{\partial \zeta^2} - \varepsilon_p \rho \frac{\partial \omega_j}{\partial t} = -\hat{r}_j \tag{3.80}$$

Boundary and initial conditions are given by:

$$\begin{aligned} &(d\omega_j/d\zeta)_{\zeta=\ell} = 0; \\ &\omega_j(0, \zeta) = \omega_{jI}(\zeta) \end{aligned} \tag{3.81}$$

The values ω_{j0} at the interface will not be known *a priori* when mass transfer resistances in the external fluid are considered. The alternative condition at $\zeta = 0$ is established by assuming known values ω_{jF} in the bulk of the fluid (eventually, as functions of time) and expressing the fluxes G_{j0} at $\zeta = 0$ in terms of mass transfer coefficients $\hat{\alpha}_j$ on the fluid side:

$$G_{j0} = \hat{\alpha}_j(\omega_{jF} - \omega_{j0})$$

On the slab side, $G_{j0} = -\rho\mathfrak{D}_{j,ef}(\partial\omega_j/\partial\zeta)_0$. Then,

$$-\rho\mathfrak{D}_{j,ef}(\partial\omega_j/\partial\zeta)_0 = \hat{\alpha}_j(\omega_{jF} - \omega_{j0}) \tag{3.82}$$

The coefficients $\hat{\alpha}_j$ are specific for each species A_j, as they significantly depend on the molecular diffusivities $\mathfrak{D}_j$.

The formulation for the component variables is obtained from Eqs. (3.80)–(3.82) by proceeding to combine them in the same way as in the main text. The results are expressed in the following compact way ($j > K$):

$$\rho\frac{\partial^2\mathbb{J}_j}{\partial\zeta^2} - \varepsilon_p\rho\frac{\partial\Omega_j}{\partial t} = 0; \tag{3.83}$$

$$\begin{aligned} &(d\mathbb{J}_j/d\zeta)_{\zeta=\ell} = 0;\\ &-\rho(\partial\mathbb{J}_j/\partial\zeta)_0 = \mathbb{A}_{jF} - \mathbb{A}_{j0};\\ &\Omega_j(0,\zeta) = \Omega_{jI}(\zeta) \end{aligned} \tag{3.84}$$

The variables $\mathbb{J}_j = \mathfrak{D}_{j,ef}\omega_j - \sum_k \hat{\sigma}_{kj}\mathfrak{D}_{k,ef}\omega_k$ were defined in Sect. 3.5 (Eq. 3.32), while the *transfer component variables* are defined as:

$$\mathbb{A}_j = \hat{\alpha}_j\omega_j - \sum_k \hat{\sigma}_{kj}\hat{\alpha}_j\omega_k; \qquad j > K \tag{3.85}$$

For a practical use of Eqs. (3.83) and (3.84), they should be written in terms of only one component variable. The following relationships arise between them from their definitions:

$$\Omega_j = \frac{\mathbb{J}_j}{\mathfrak{D}_{j,ef}} - \sum_k \hat{\sigma}_{kj}\left(1 - \frac{\mathfrak{D}_{k,ef}}{\mathfrak{D}_{j,ef}}\right)\omega_k \tag{3.86}$$

$$\frac{\mathbb{A}_j}{\hat{\alpha}_j} = \frac{\mathbb{J}_j}{\mathfrak{D}_{j,ef}} - \sum_k \hat{\sigma}_{kj}\left(\frac{\hat{\alpha}_k}{\hat{\alpha}_j} - \frac{\mathfrak{D}_{k,ef}}{\mathfrak{D}_{j,ef}}\right)\omega_k \tag{3.87}$$

$$\frac{\mathbb{A}_j}{\hat{\alpha}_j} = \Omega_j + \sum_k \hat{\sigma}_{kj}\left(1 - \frac{\hat{\alpha}_k}{\hat{\alpha}_j}\right)\omega_k \tag{3.88}$$

The choice of variable $\mathbb{J}_j$ is the best suited, as the effect of the $\mathbb{A}_j$ in Eqs. (3.86)–(3.88) vanishes at very high values of $\hat{\alpha}_j$ (i.e., when the external resistances become negligible), and the effect of the Ω_j vanishes in SS ($\partial\Omega_j/\partial t = 0$).

Considering a transient state, once Ω_j and $\mathbb{A}_j$ from Eqs. (3.86) and (3.87) are replaced in Eqs. (3.83) and (3.84), it can be concluded that the solution for $\mathbb{J}_j$ will in general depend on the mass fractions ω_k, because of ratios $\hat{\alpha}_k/\hat{\alpha}_j$ and $\mathfrak{D}_{k,ef}/\mathfrak{D}_{j,ef}$ different to one. Therefore, variables $\mathbb{J}_j$ will not behave as reaction invariants, probably in a significant degree. The same conclusion can be reached for variables Ω_j and $\mathbb{A}_j$.

In steady state, Eqs. (3.83) and (3.84) become:

$$\rho \frac{\partial^2 \mathbb{J}_j}{\partial \zeta^2} = 0; \tag{3.89}$$

$$\begin{aligned} &(d\mathbb{J}_j/d\zeta)_{\zeta=\ell} = 0; \\ &-\rho\left(\frac{\partial \mathbb{J}_j}{\partial \zeta}\right)_0 = \hat{\alpha}_j[\frac{\mathbb{J}_{jF}-\mathbb{J}_{j0}}{\mathfrak{D}_{j,ef}} - \sum_k \hat{\sigma}_{kj}\left(\frac{\hat{\alpha}_k}{\hat{\alpha}_j} - \frac{\mathfrak{D}_{k,ef}}{\mathfrak{D}_{j,ef}}\right)(\omega_{kF} - \omega_{k0})] \end{aligned} \tag{3.90}$$

From Eq. (3.89), the variables $\mathbb{J}_j$ remain uniform:

$$\mathbb{J}_j = \mathbb{J}_{j0}; \qquad 0 \le \zeta \le \ell \tag{3.91}$$

Equation (3.90) is reduced to:

$$\frac{\mathbb{J}_{jF} - \mathbb{J}_{j0}}{\mathfrak{D}_{j,ef}} = \sum_k \hat{\sigma}_{kj}\left(\frac{\hat{\alpha}_k}{\hat{\alpha}_j} - \frac{\mathfrak{D}_{k,ef}}{\mathfrak{D}_{j,ef}}\right)(\omega_{kF} - \omega_{k0}) \tag{3.92}$$

Equation (3.92) indicates that in general the values of $\mathbb{J}_{j0}$ will depend on the differences $(\omega_{kF} - \omega_{k0})$, which in turn will depend on the reaction rates inside the pellets. Hence, the variables $\mathbb{J}_{j0}$ will not behave as reaction invariants when external resistances are considered. However, the use of these variables will be still helpful in the numerical resolution of the problem. To see this assertion, the SS balances of the key species are first written. From Eqs. (3.80) to (3.82):

$$\rho \mathfrak{D}_{k,ef} \frac{\partial^2 \omega_k}{\partial \zeta^2} = -\hat{r}_k; \qquad k = 1, \ldots, K \tag{3.93}$$

$$\begin{aligned} &(d\omega_k/d\zeta)_{\zeta=\ell} = 0; \\ &-\rho \mathfrak{D}_{k,ef}(\partial \omega_k/\partial \zeta)_0 = \hat{\alpha}_k(\omega_{kF} - \omega_{k0}) \end{aligned} \tag{3.94}$$

By definition, $\mathfrak{D}_{j,ef}\omega_j = \mathbb{J}_j + \sum_k \hat{\sigma}_{kj}\mathfrak{D}_{k,ef}\omega_k$ (Eq. 3.34 in the main text). Employing Eqs. (3.91) and (3.92), this expression can be rewritten as:

$$\begin{aligned} \omega_j(\zeta) = \frac{\mathbb{J}_{jF}}{\mathfrak{D}_{j,ef}} - \sum_k \hat{\sigma}_{kj}\left(\frac{\hat{\alpha}_k}{\hat{\alpha}_j} - \frac{\mathfrak{D}_{k,ef}}{\mathfrak{D}_{j,ef}}\right)(\omega_{kF} - \omega_{k0}) \\ + \sum_k \hat{\sigma}_{kj} \frac{\mathfrak{D}_{k,ef}}{\mathfrak{D}_{j,ef}} \omega_k(\zeta) \end{aligned}; \qquad j > K \tag{3.95}$$

Then, consider that Eqs. (3.93) and (3.94) has been discretised to evaluate iteratively the nodal values of ω_k, including ω_{k0}. In each node, the fractions ω_j of the component species, needed to evaluate the rates $\hat{r}_k$, can be readily evaluated from Eq. (3.95) by using the current values of ω_k in the same node. Then, the dimension of the numerical problem is effectively reduced to the number K of key species.

Appendix 4: Evaluation of Reaction Rates in Turbulent Regime

Time-smoothed variables will be here enclosed by angle-brackets "$\langle\rangle$", while normal symbols are used for instantaneous values. In particular, $\langle\hat{r}_j\rangle$ is time-smoothed net mass-production rate of A_j and $\hat{r}_j$ is the instantaneous value. Note that no distinction was made in Sect. 3.6.

A broad variety of concepts and methods have been proposed to estimate the rates $\langle\hat{r}_j\rangle$. A specific approach will be presented here with the purpose of appreciating the additional complexity involved in the evaluation of reaction rates under turbulent conditions. The chosen method corresponds to an alternative available in the platform ANSYS-FLUENT [21], suggested for the formation rates of NO_X, SO_X and soot in turbulent combustion.

As an elementary example, consider that the net formation rate of a species A_j is given by $\hat{r}_j = -kC_j^2$, where k is the kinetic coefficient. Expressing $C_j = \rho\omega_j/m_j$, $\hat{r}_j = -k(\rho/m_j)^2\omega_j^2$. The value $\langle\hat{r}_j\rangle$ in a given position of a turbulent flow reactor, assuming that density fluctuations can be neglected ($\langle\rho\rangle = \rho$) and that the reaction proceeds under isothermal conditions, is expressed as $\langle\hat{r}_j\rangle = -k(\rho/m_j)^2\langle\omega_j^2\rangle$. Therefore, the evaluation of $\langle\omega_j^2\rangle$ is required.

The method here discussed is based on the use of the *probability density function* (PDF) of a given variable ϕ, denoted by $\mathcal{P}(\phi)$. Accordingly, $[\mathcal{P}(\phi)d\phi]$ is the probability that at a given time in a given position the variable takes values between ϕ and $(\phi + d\phi)$. The PDF should verify $\int_\phi \mathcal{P}(\phi)d\phi = 1$, where the integration is carried out on the range of feasible values of ϕ. Then, for a function $f(\phi)$, the time-smoothed value is evaluated as:

$$\langle f(\phi)\rangle = \int_\phi f(\phi)\mathcal{P}(\phi)d\phi$$

For example, with $\phi \equiv \omega_j$:

$$\langle\hat{r}_j\rangle = -k(\rho/m_j)^2\int_0^1 \omega_j^2\mathcal{P}(\omega_j)d\omega_j \tag{3.96}$$

In general, $\hat{r}_j$ will depend on more than one state variable, $\hat{r}_j = \hat{r}_j(\phi_1, \phi_2, \ldots)$. Strictly, it should be considered a *joint*-PDF, $\mathcal{P}(\phi_1, \phi_2, \ldots)$, for the probability of the simultaneous occurrence of values between $[\phi_1, \phi_1 + d\phi_1]$, $[\phi_2, \phi_2 + d\phi_2]$, and so on. An important simplification in the ANSYS-FLUENT method is assuming that the events of each variable are independent of the remainder variables, i.e., $\mathcal{P}(\phi_1, \phi_2, \ldots) = \mathcal{P}(\phi_1)\mathcal{P}(\phi_2)\ldots$. Then, it can be written:

$$\langle \hat{r}_j \rangle = \int_{\phi_1,\phi_2,\ldots} \hat{r}_j(\phi_1, \phi_2, \ldots)\mathcal{P}(\phi_1)\mathcal{P}(\phi_2)\ldots d\phi_1 d\phi_2 \ldots \tag{3.97}$$

As another example, if the consumption of a species A proceeds in the presence of the inhibitor species B so that $\hat{r}_A = -c\ e^{-E/(RT)}\omega_A^{1/2}/(1 + K\omega_B)$, with constants c, E/R, K:

$$\langle \hat{r}_A \rangle = -c \int_{T,\omega_A,\omega_B} e^{-E/(RT)} \frac{\omega_A^{1/2}}{1 + K\omega_B} \mathcal{P}(T)\mathcal{P}(\omega_A)\mathcal{P}(\omega_B) dT d\omega_A d\omega_B$$

Regarding $\mathcal{P}(\phi)$, the *two-moment beta function* is usually employed for expressing $\mathcal{P}(\phi)$:

$$\mathcal{P}(\phi) = \frac{\Gamma(\alpha+\beta)}{\Gamma(\alpha)\Gamma(\beta)}\phi^{\alpha-1}(1-\phi)^{\beta-1} = \frac{\phi^{\alpha-1}(1-\phi)^{\beta-1}}{\int_0^1 \phi^{\alpha-1}(1-\phi)^{\beta-1} d\phi} \tag{3.98}$$

where Γ is the Gamma function and parameters α and β are defined in terms of the time-smoothed value $\langle\phi\rangle$ and variance σ_ϕ^2:

$$\alpha = \langle\phi\rangle\left[\frac{\langle\phi\rangle(1-\langle\phi\rangle)}{\sigma_\phi^2} - 1\right]; \qquad \beta = (1-\langle\phi\rangle)\left[\frac{\langle\phi\rangle(1-\langle\phi\rangle)}{\sigma_\phi^2} - 1\right] \tag{3.99}$$

The *two-moment beta function* requires that the range of ϕ is normalised between 0 and 1. In this way, it can be easily shown that the consistency relation $\langle\phi\rangle = \int_0^1 \phi\mathcal{P}(\phi)d\phi$ is verified, independently of the value σ_ϕ^2.

In a non-isothermal reactor, one of the variables should be the temperature, as the kinetic coefficients strongly depend on T. Then, lower and upper bounds in a given problem must be estimated, T_m and T_M, which allow defining the variable $\phi_T = (T - T_m)/(T_M - T_m)$.

To evaluate the variance σ_ϕ^2, ANSYS-FLUENT uses the following expression, associated with the *κ-ε turbulence model*:

$$\sigma_\phi^2 = \frac{\mu^t \kappa c_g}{\rho \varepsilon c_d}\left[\left(\frac{\partial\langle\phi\rangle}{\partial z_1}\right)^2 + \left(\frac{\partial\langle\phi\rangle}{\partial z_2}\right)^2 + \left(\frac{\partial\langle\phi\rangle}{\partial z_3}\right)^2\right], \tag{3.100}$$

where z_i are cartesian coordinates, c_g and c_d are fitting coefficients, $c_g = 2.86$, $c_d = 2$, μ^t is the turbulent viscosity, κ (m^2/s^2) is the *turbulent kinetic energy* and ε (m^2/s^3) the *dissipation rate*. The quantities μ^t, κ and ε are local values available after solving the equations of momentum and energy for the fields of time-smoothed velocity and pressure.

As the foregoing formulation is an approximation, it should be controlled that $\mathcal{P}(\phi)$ defined by Eqs. (3.98, 3.99) yields acceptable results. If not, $\mathcal{P}(\phi)$ may be extremely wide or it can even be bimodal. Thus, the value σ_ϕ should be restrained in a way to avoid values of the parameters α and β from Eq. (3.99) lower than 1. This is warrantied if $\sigma_\phi < \langle\phi\rangle(1-\langle\phi\rangle)$. The condition $\sigma_\phi \leq 0.9\langle\phi\rangle(1-\langle\phi\rangle)$ can be practically used.

References

1. Santa Cruz, J. A., Mussati, S. F., Scenna, N. J., Gernaey, K. V., & Mussati, M. C. (2016). Reaction invariant-based reduction of the activated sludge model ASM1 for batch applications. *Journal of Environmental Chemical Engineering, 4*(3), 3654–3664.
2. Fjeld, M. (2018). Model reduction in large chemical systems—An alternative method applying discrete approximation. *ChemRxiv*. https://doi.org/10.26434/chemrxiv.7284932.v1
3. Rodrigues, D., Billeter, J., & Bonvin, D. (2017). Generalization of the concept of extents to distributed reaction systems. *Chemical Engineering Science, 171*, 558–575.
4. Billeter, J., Rodrigues, D., Srinivasan, S., Amrhein, M., & Bonvin, D. (2018). On decoupling rate processes in chemical reaction systems—Methods and applications. *Computers and Chemical Engineering, 114*, 296–305.
5. Bird, R. B., Stewart, W. E., Lightfoot, E. N., & Klingenberg, D. J. (2015). *Introductory transport phenomena*. Wiley.
6. Levenspiel, O. (1996). *The chemical reactor omnibook* (p. 64.18). Oregon st. University Book Stores.
7. Wilke, C. R. (1950). Diffusional properties of multicomponent gases. *Chemical Engineering Progress, 46*, 95–104.
8. Kee, R. J., Coltrin, M. E., & Glarborg, P. (2003). *Chemically reacting flows*. Wiley-Interscience.
9. Green, D. W., & Perry, R. H. (Eds.) (2008). *Perry's chemical engineers' handbook* (8th ed.). The McGraw-Hill Companies, Inc.
10. Levenspiel, O. (1998). *Chemical reaction engineering* (3rd ed.). Wiley.
11. Jurtz, N., Kraume, M., & Wehinger, G. D. (2019). Advances in fixed-bed reactor modeling using particle-resolved computational fluid dynamics (CFD). *Reviews in Chemical Engineering, 35*(2), 139–190.
12. Jiménez-Gómez, C. P., Cecilia, J. A., Durán-Martín, D., Moreno-Tost, R., Santamaría-González, J., Mérida-Robles, J., Mariscal, R., & Maireles-Torres, P. (2016). Gas-phase hydrogenation of furfural to furfuryl alcohol over Cu/ZnO catalysts. *Journal of Catalysis, 336*, 107–115.
13. Kaza, K. R., & Jackson, R. (1980). Diffusion and reaction of multicomponent gas mixtures in isothermal porous catalyst. *Chemical Engineering Science, 35*, 1179–1877.
14. Gualtieri, C., Angeloudis, A., Bombardelli, F., Jha, S., & Stoesser, T. (2017). On the values for the turbulent schmidt number in environmental flows. *Fluids, 2*(17).
15. Feng, Y., Qin, J., Zhang, S., Bao, W., Cao, Y., & Huang, H. (2015). Modeling and analysis of heat and mass transfers of supercritical hydrocarbon fuel with pyrolysis in mini-channel. *International Journal of Heat and Mass Transfer, 91*, 520–531.
16. Luzi, C. D., Mariani, N. J., Asensio, D. A., Martínez, O. M., & Barreto, G. F. (2019). Estimation of the radial distribution of axial velocities in fixed beds of spherical packing. *Chemical Engineering Research and Design, 150*, 153–168.
17. Delgado, J. M. P. Q. (2007). Longitudinal and transverse dispersion in porous media. *Chemical Engineering Research and Design, 85*, 1245–1252.
18. Froment, G. F., Bischoff, K. B., De Wilde, J. (2011). *Chemical reactor analysis and design* (3rd ed.). Wiley.
19. Hosseini, S., Moghaddas, H., Soltani, S. M., & Kheawhom, S. (2020). Technological applications of honeycomb monoliths in environmental processes: A review. *Process Safety and Environmental Protection, 133*, 286–300.
20. Dixon, A.G., & Partopour, B. (2020). Computational Fluid Dynamics for fixed bed reactor design. *Annual Review of Chemical and Biomolecular Engineering, 11*(1), 109–130. https://doi.org/10.1146/annurev-chembioeng-092319-075328
21. Ansys® Fluent Theory Guide, Release 2021 R1, Chap. 14, ANSYS, Inc.

Printed in the United States
by Baker & Taylor Publisher Services